WOODLANDS IN CRISIS

A Legacy of Lost Biodiversity on the Colorado Plateau

WOODLANDS IN CRISIS

A Legacy of Lost Biodiversity on the Colorado Plateau

by

Gary Paul Nabhan
Marcelle Coder
Susan J. Smith

with contributions by
Zsuzsi I. Kovacs

With support from the
National Commission on
Science for Sustainable Forestry

BILBY RESEARCH CENTER OCCASIONAL PAPERS NO. 2

ISBN 0-9718786-6-8

ACKNOWLEDGMENTS

This report was funded by the National Commission on Science and Sustainable Forestry, with Norm Christensen as our liaison. We would like to express our gratitude to reviewers Lisa Floyd, Peter Friederici, and Steve Buckley. We thank Louella Holter, series editor, for extraordinary editorial assistance; Dan Boone, Ron Redsteer, and Patrick McDonald for graphics, and Bruce Hooper for archival research assistance (all of the Bilby Research Center, NAU). We further thank Gwenn Gallenstein of the National Park Service; NAU colleagues Catherine Gehring, Department of Biological Sciences; Wally Covington, Margaret Moore, Pete Fulé, Diane Vosick, Judy Springer, Peter Friederici, and Lisa Machina of the Ecological Research Institute; Thomas Whitham and Neil Cobb of the Merriam-Powell Center for Ecological Research; Tom Sisk and Scott Anderson of the Center for Environmental Science and Education; Ken Cole, USGS Colorado Plateau Research Station; Ron Hiebert, National Park Service/Cooperative Ecosystems Studies Unit, for making available important information from the Phase I report on monitoring in the southern Colorado Plateau parks; and Matthew Clark for insights on predator control. In addition we thank Craig Allen, Brian Jacobs, John Mack, Julio Betancourt, Randy Balice, Brad Vierra, Lisa Floyd, Marilyn Colyer, Carolyn Landes, Liz Bauer, George San Miguel, Randy Travis, Terra Travis, Joel Glansberg, Melissa Savage, Paul Whitefield, John Brinkley, Terralene Foxx, Brett Dickson, and Don Falk for site-specific data.

THE RALPH M. BILBY RESEARCH CENTER

The Ralph M. Bilby Research Center was founded in 1982 to promote interdisciplinary research at Northern Arizona University. The Bilby Research Center has been affiliated with the Center for Sustainable Environments since 1999, when the CSE was founded. The Bilby Occasional Papers Series provides an outlet for quality research on a wide range of topics. For more information please contact us at Box 6013, Northern Arizona University, Flagstaff, 86011.

CONTENTS

List of Tables   *ix*

List of Figures   *x*

Woodlands in Crisis   *1*

The Reference Envelope   *5*

    Geographic Context   *5*

    Fire History   *9*

    Livestock History   *9*

    Introduction of Sheep into the West   *13*

    Logging and Shrub Removal   *13*

    Indigenous Fire Management   *17*

    Faunal (Avian) Diversity   *21*

    Predator Control   *23*

    Insect Outbreaks   *24*

    Climate Change   *27*

    The Forest, Climate, Fire, and Insect Connection   *33*

Four Case Studies from the Colorado Plateau   *35*

    The Jemez Mountains and Bandelier National Monument   *35*

    Mesa Verde National Park   *45*

    The Chuska Mountain Complex   *59*

    The San Francisco Volcanic Field   *69*

    Summary of Case Studies   *78*

Implications for Management and Restoration   *81*

    Global Climate Change   *81*

    Traditional Ecological Knowledge   *83*

    Invasive Plant Species   *84*

    Conclusions and Recommendations   *87*

Literature Cited   *91*

Appendix A: Invasive Plant Species from Our Study Areas

Appendix B: Rare, Threatened, and Endangered Plants from Our Study Areas

LIST OF TABLES

1. Sources of information from the Colorado Plateau and suggested causes of change — 7

2. Asynchrony in fire frequency decline in southwestern ponderosa pine forests — 10

3. Fire-stimulated plants — 18

4. Use of fire among three broad cultural subsistence groups on the Colorado Plateau — 20

5. Hypothesized avian population changes from pre-1850 conditions, inferred from survey plots — 22

6. Forest tree insects — 25

7. Strengths and limitations of paleoecological proxy data sources — 28

8. Major climate intervals from El Malpais tree-ring record — 31

9. Plant species recovered from archaeological sites in the eastern Jemez Mountains — 39

10. Plant uses and numbers of plant species used from Jemez Mountain plant communities — 40

11. Plant species recovered from archaeological sites in the Mesa Verde region — 51

12. Plant species recovered from archaeological sites in the Chuska Mountain region — 62

13. Plant species recovered from archaeological sites in the Wupatki and Sunset Crater National Monuments — 73

14. Current threats in our study areas, as identified by land managers — 79

LIST OF FIGURES

1. Cattle round-up in Moencopi Wash near Tuba City, Arizona, 1941   11

2. Sheep crossing the Rio Grande River on the Buckman Bridge, 1922   14

3. Arizona Lumber and Timber Company sawmill, 1899   15

4. Logging camp south of Flagstaff, 1904   16

5. Logging camp south of Flagstaff, 1904   17

6. Average annual precipitation recorded at 97 weather stations from 1900 to 2000   32

7. Colorado Plateau showing locations of our four study areas   36

8. The Jemez Mountains and Bandelier National Monument   37

9. Aerial photograph of fields in the Los Alamos area   41

10. Mesa Verde National Park   46

11. Mesa Verde cuesta, Colorado   47

12. Repeat photography in Prater Canyon, near Prater homestead and windmill   50

13. Brush burning on reservation, east of Mancos River. September 26, 1960   54

14. Range burned by Utes, Mancos Canyon, July 1933   55

15. The Chuska Mountains and Canyon de Chelly National Monument   59

16. Canyon de Chelly National Monument   60

17. San Francisco Volcanic Field   69

18. Wupatki and Sunset Crater National Monuments   70

WOODLANDS IN CRISIS:

A Legacy of Lost Biodiversity on the Colorado Plateau

The greatest challenge in assembling and interpreting a land-use history of the Southwest is disentangling cultural from natural causes of environmental change. (Allen et al. 1998:71)

In recent years, the American West has suffered from unprecedented stand-replacing wildfires, and the government has invested more money in preventative forest thinning than ever before. This crisis for western forests and woodlands has spurred heated policy debates about thinning, controlled burns, and other restoration activities. Activists, loggers, biologists, and recreationists on the Colorado Plateau differ in their opinions regarding the degree to which thinning or controlled burning can truly serve to restore wooded habitats, and what reference conditions or restoration goals are needed to guide such plans. One proposed tool for guiding restoration is to use "pre-settlement conditions" to understand what forest stand structure was like before European settlement, and to use restored stand structure to drive restored ecosystem processes, such as frequent, low-intensity fire (Covington and Moore 1994). Land managers who advocate thinning may assume that thinning alone is enough to restore process and health to forests and woodlands, whereas opponents have assumed that resource managers believe that restoration ends with thinning (P. Friederici, personal communication 2004). Further complicating the issue is that policy-makers—who may have different goals—have taken parts of the message and attempted to apply a one-size-fits-all forest thinning policy across the West's heterogeneous landscapes (Healthy Forests 2002).

There is no doubt that an understanding of the shapers of our modern forests and woodlands is an important tool for restoration, and understanding stand structure and past fire frequencies is an important part of that picture. However, there are caveats associated with using presettlement conditions as the sole reference point for restoration (Christensen 1989; Egan and Howell 2001; Langston 1999; Swetnam et al. 1999). As will become clear through our use of case studies, cultural modifications of the landscape began well before European settlement of the West, and these important drivers of change must be taken into account; use of a one-point-in-time referent can miss these changing cultural conditions. Also, it may be difficult or impossible to know with certainty what these conditions actually were. A generalized set of structural conditions may not apply to specific local environments. Using structure alone—or even in combination with fire frequency—to predict ecosystem function will likely miss such important pieces of the ecosystem as the role and importance of biodiversity and understory composition, effects of the understory on hydrology, changes in nutrient cycling dynamics, the role of predators, birds, and other wildlife, and the importance of non-timber forest products. Finally, using past conditions as a model to restore modern forests also underestimates or ignores such new influences as pollution, habitat frag-

Following the Sand Point Fire, Lewis and Clark National Forest, July 1985
Photo courtesy of Northern Arizona University

mentation and loss, exotic invasive species, and climate change. As we shall see, restoration plans based on such goals may be able to replicate earlier stand structures, but may nevertheless fall short of addressing resilience, ecosystem processes such as hydrology and decadal climate fluctuation, modern influences on forest and woodland ecology, and the ecological and cultural implications of thinning.

Here we evaluate the extent to which the goal of restoration to presettlement conditions that has developed around Flagstaff, Arizona (as outlined by authors in Friederici 2003b) is applicable and transferable to other pine-dominated landscapes in the western United States. We are particularly concerned with the use of this model to retain, restore, or enhance native biodiversity in these landscapes while attempting at the same time to reduce the probability of high-severity

crown fires, or manage forests and woodlands for commodity production. As an alternative, we offer a primer for understanding how diverse land-use histories have impacted the health of pine-dominated ecosystems in the West, and we offer measures for better managing them in the future. We draw on a systematic review of the historic effects of land use and climate on ecosystem health, biodiversity, and non-timber forest products in four specific landscapes on the Colorado Plateau—the Jemez Mountains in New Mexico, Mesa Verde in Colorado, the Chuska Mountains in Arizona, and the San Francisco Volcanic Field in Arizona—all of which have long histories of human occupation and use. In these landscapes, we evaluate the degree to which livestock grazing, fire suppression, and other practices have changed the frequency, severity, and extent of fires, the species richness of understory plants, and the availability of non-timber forest products formerly harvested by Native, Hispanic, and Anglo American communities.

One stated goal of restoration is "the preservation of biodiversity, the health and maintenance of sustainable ecosystems, and the development of mutually beneficial relationships between humans and nature" (Friederici 2003a:3). Resource managers and researchers generally agree that ecological restoration of these forests and woodlands to more "natural" conditions is urgently needed (Allen et al. 2002; Covington et al. 1997; Dahms and Geils 1997). This consensus has developed from a growing body of literature that documents how more frequent, intense, and large wildfires—as well as the resulting fire suppression and rehabilitation costs—have been increasing in the ponderosa pine forests of Arizona and New Mexico over the last decade. Fires that have ignited naturally, accidentally, or intentionally in ponderosa pine forests have also been spreading more frequently to the mixed conifer forests above and pinyon-juniper woodlands below the ponderosa zone. Between 1993 and 2002 there was a steady increase in annual acreage burned, while the costs of fire suppression and reforestation quadrupled (USDA Forest Service 2002). These changes in fire dynamics have been related to historic changes in the density and age structure of ponderosa pine forests (Covington and Moore 1994) and of some pinyon-juniper woodlands (Baker and Shinneman 2004). Nevertheless, scientists differ in the degree to which they attribute these structural changes in vegetation to fire exclusion and suppression policies, livestock grazing, climatic fluctuations, bark beetle infestations, or other factors (Baker and Shinneman 2004; Dahms and Geils 1997; Floyd et al. 2004).

Disagreements with regard to the causes of these problematic changes—as well as the relative efficacy of potential solutions—are no longer restricted to the realm of academic debate. The debate has spread to concern society as a whole, as indicated by heated discussions and intense media coverage prior to the passage of Healthy Forests legislation by the U.S. Congress in the fall of 2003 (Healthy Forests 2002). There is little doubt that there will soon be more federal government support for addressing restoration and fire management issues than ever before in American history. The central question, then, has become this: How can this unprecedented influx of support for forest management be used not merely to reduce the frequency of property-damaging fires over the short term, but also to restore the health, resilience, and diversity of forests and woodlands over the long term?

Unfortunately, there remains considerable scientific uncertainty about how to best address these challenges in ponderosa pine forests (Allen et al. 2002; Dahms and Geils 1997; Kloor 2000), not to mention in mixed conifer forests and pinyon-juniper woodlands (Baker and Shinneman 2004; Floyd 2003; Floyd et al. 2004). Moreover, there has been intense controversy among environmental activists, forestry professionals, and other stakeholders who do not agree on the relative economic, ecologic, and aesthetic costs and benefits of the various practices being championed as ways to restore these forests and woodlands (Allen et al. 2002; Friederici 2003a; Jenkins 2001; Nijhuis 1999).

When dealing with somewhat comparable issues where the economic stakes and environmental or human health risks are high—issues such as reducing the impacts of catastrophic floods, droughts, food-borne illnesses, and disease epidemics—systematic reviews have been adopted as essential means for reducing scientific uncertainty (Altman et al. 2001; Baker and Shinneman 2004; Nabhan 2003). Such systematic reviews are especially helpful when isolated case studies on how to best reduce risks have produced inconsistent or apparently contradictory results (Altman et al. 2001). In this case the stakes are definitely high, whereas scientific certainty about ultimate causes and best practices remains low, demonstrating a clear need for a systematic review of management options affecting the biodiversity and health of pine-dominated ecosystems. The value of such a review has been recently demonstrated for a subset of these issues by Baker and Shinneman (2004), who reviewed 70 potential studies for their insights into seven questions regarding fire and restoration in pinyon-juniper woodlands in 11 western states. They found that 46 of the studies met their criteria well enough to help discern fire ecology trends, and to help propose restoration options for some (but not all) of the varying plant associations in which pinyon pines and junipers are found.

We wish to underscore the need for a similar review of the effects of land-use history on the biodiversity of pine-dominated ecosystems in the western United States. The potential economic, ecologic, and health costs of inappropriate management actions in these ecosystems—or no action at all—have never been higher. An analysis of the 2002 Rodeo-Chedeski fire in east-central Arizona determined that decades of the "no restoration" treatment method resulted in costs of $300 million for fire-fighting, property losses, and long-term rehabilitation of watersheds through reforestation and grass reseeding (Snider et al. 2003). The 467,000 acres of burned tribal, Forest Service, state, and private lands cost $582 per acre, com-

pared to the preliminary estimate of $505 per acre for forest restoration treatments that might effectively prevent catastrophic fires and avoid associated fire suppression costs. Nevertheless, such restoration efforts have to date occurred on only a miniscule portion of lands dominated by pines in the 11 western states. Each year that no action is taken to restore these forests and woodlands to healthier conditions increases the percentage classified as Class 3 (highest risk) for "unnatural fires." Just nine of the western states already have 12 million acres of pine forest in Class 3 "high risk" condition. For this acreage—not including forested acreage that may soon achieve Class 3 status, as well as all "at risk" pinyon-juniper woodlands—it would now be more cost effective to invest $6 billion in restoring the health of these ecosystems than to spend an equivalent amount to fight future fires and pay for property losses should these attempts fail (Snider et al. 2003). In short, we believe that the investment by the National Commission on Science for Sustainable Forestry in our efforts toward a systematic review of biodiversity and ecosystem health issues in forest and woodland landscapes could not have been more timely.

THE REFERENCE ENVELOPE

In the rapidly expanding primary literature, scientists and historians have cited a wide variety of causes that they think have contributed to historic changes in fire dynamics, vegetation structure, biodiversity, and ecosystem health in pine-dominated landscapes of the Colorado Plateau (Table 1). In a systematic review of this list of potential causes of change, three emerging patterns stand out: (1) Certain anthropogenic changes in pine-dominated landscapes began well before European settlement of particular forests and woodlands, and occurred asynchronously across the Colorado Plateau ecoregion. (2) Contrary to some accounts, fire frequencies also began to decline asynchronously, but in most cases well in advance of the formal fire suppression policy implemented by government agencies such as the USDA Forest Service and the USDI Bureau of Indian Affairs. (3) The severity of change is not only habitat specific—varying with the soils, climate, flora, and relative geographic isolation of a particular site—but it has also varied greatly among landscapes managed by the 23 native and immigrant cultures of the Colorado Plateau (Nabhan et al. 2002a, 2002c).

Variations in these patterns suggest that there can be no single presettlement reference point that predates all significant human causal factors, or that could serve in a straightforward manner to guide all restoration efforts in the region. Instead, managers interested in restoration must attempt to determine a "reference envelope" (Allen et al. 2002; Egan and Howell 2001; D. Falk, personal communication 2003; Savage 2003) rather than seeking a single reference point.

This reference envelope should capture, through various methods of historical ecology, the historic range of variation of conditions in any ecosystem, rather than simply determining the earliest recorded pre-European settlement condition or even the earliest pre–Native American condition (Egan and Howell 2001; Savage 2003). In the following review of potential causal factors we attempt to determine the range of historic and geographic variation for each factor, and then we consider the relative importance of each factor, relative to others that have been proposed.

Geographic Context

Forests dominated by ponderosa pines and closely related species cover some 30 million acres in the western United States, adjacent to Canada and northern Mexico (Covington 2003; Little 1971). Pygmy woodlands dominated by two species of pinyon pines and four species of juniper cover at least 15 million additional acres in the Southwest (Van Hooser et al. 1993). This vegetation type represents 35.5 percent of the Colorado Plateau, and is the third largest vegetation type in the United States (Rundall et al. 2003), but because it variously grades into sagebrush scrub, savanna, and grassland, its precise extent is difficult to quantify (Mitchell and Roberts 1999). There is great variation in elevational amplitude for both of these vegetation types, depending on soils, local microclimate, and aspect. The highest reaches of ponderosa pines occur at about 9500 ft and the lowest reaches of pinyon-juniper at about 3000 ft (Covington 2003; Nabhan and Wilson 1995). We focus on three major vegetation types within this range of pine-dominated ecosystems, each encompassing different biotic communities.

Mixed conifer forests occur at high elevations in the mountains on the Colorado Plateau, and to some extent on the rims and in the drainages of canyon country. They are dominated not only by ponderosa pine but also by Douglas fir (*Pseudotsuga menziesii*), white fir (*Abies concolor*), and blue spruce (*Picea pungens*). Clonal patches of aspen (*Populus tremuloides*) and Gambel oak (*Quercus gambelii*) occur on slopes disturbed by fire or other dynamic landscape processes. Numerous grasses, herbs, and fungi may occur beneath the more open canopies of these forests, but except for fungi, understory species may be sparse under the more moist, closed-canopy stands in canyons and ravines (Mac 1998). One prevailing view of mixed conifer forest dynamics is that mean fire intervals were decadal in scale until the late 1880s, when widespread fires stopped in some places. Ponderosa pines, once co-dominant in many stands, have been replaced by dense young understories of Douglas fir and white fir. Forest stand inventories suggest that between 1962 and 1986, there was an 81 percent increase in the area of stands classified as mixed conifer forests as opposed to ponderosa pine forests. Herbaceous understories have been reduced by denser canopies and needle litter, with heavier surface fuels resulting in increased risk of crown fires (Mac 1998).

Ponderosa pine forests, which grade into mixed conifer forests at higher elevations and into pinyon-juniper woodlands or park-like grasslands or sage scrub below, are found in at least 21 habitat types in the Southwest (Allen et al. 2002). They occur on the slopes of mountains on the Colorado Plateau and adjacent sky islands as well as in more gentle terrain, including plateaus, valleys, and mesas. The composition of these various forest habitat types has been differentially shaped through time by stochastic and deterministic processes such as frequent fires, insect infestations, episodic regeneration of cohorts responding to El Niño–Southern Oscillation (ENSO) climatic cycles, and regional climatic events like droughts (Allen et al. 2002). These processes have created heterogeneous spatial patterns at both local and landscape levels. Ponderosas can form dense closed-canopy stands, or heterogeneous mosaics with shrubs, including aspen, Gambel oak, gray oak (*Quercus grisea*), and manzanita (*Arctostaphylos* sp.), and can surround park-like meadows and wetlands. Understory vegetation in open stands can be incredibly rich in herbaceous species that support diverse wildlife.

One prevailing view is that widespread surface fires occurring at 4 to 36 year intervals kept many of the ponderosa forests open and diverse until the 1880s in some places, and until 1912 in others (Covington and Moore 1994; Swetnam and Baisan 1996). Since then, doghair thickets of ponderosa saplings and their needle mats have increased in density, with old overstory trees declining or dying (Covington and Moore 1994; Mac 1998). Mature tree densities measured around 1900 by Woolsey ranged between 8 and 51 trees per acre, but in the same plots now exceed 1000 trees per acre (Allen et al. 2002; D. Falk, personal communication 2003). The ponderosa pine zone is actually shrinking as the fire-intolerant communities at the elevation limits for ponderosa expand their range—fir in the mixed conifer has moved downslope, and pinyon in the lower elevation woodland has moved upslope.

Nutrient cycling dynamics have been arrested in many places because of changes in forest structure and composition, and landscape-level biodiversity is presumed to have dramatically decreased (Allen et al. 2002). Understory grasses and forbs have decreased in abundance and species richness, replaced by mats of decomposing needles (Allen et al. 2002; Covington and Moore 1994). Herbaceous understory plants have decreased from as many as 60 species to less than 35 in one natural area (Deichmann 1980; Mac 1998). Understory plant cover at other reference sites that experience periodic low-intensity wildfires may still be as high as 37 percent, but cover has dwindled to less

Table 1. Sources of information from the Colorado Plateau and suggested causes of change.

Site	Author	Data Types	Time Period	Suggested Causes of Change
Canyon de Chelly	Betancourt & Davis 1984	Pack rat middens	11,900 BP to present	Climate variability
Canyon de Chelly	Fall et al. 1981	Pollen	Unknown	Climate, vegetation change
Canyon de Chelly	Reed & Hensler 1999	Archaeological (pollen, macros, artifacts)	AD 500–1200	—
Chaco Canyon	Floyd et al. 2003c	Vegetation plots	AD 1948 to present	Grazing excluded
Colorado Plateau	Anderson 1989	Pollen, packrat middens	14,000 BP to present	Climate variability
Colorado Plateau	Betancourt et al. 1990	Packrat middens	40,000 BP to present	Climate, biogeography
Colorado Plateau	Swetanam & Betancourt 1990	Tree-ring fire scar study	AD 500 to present	Climate, fire
Chuska Mountains	Wright et al. 1973	Pollen	11,000 BP to present	Climate, vegetation change
Jemez Mountains	Allen & Breshears 1998	Repeat aerial photography	AD 1935–1975	Drought
Jemez Mountains	Brunner-Jass 1999	Bog core study (pollen and & charcoal)	11,800 BP to present	Fire, vegetation, and climate history
Jemez Mountains	Grissino-Mayer & Swetnam 1997	Tree-ring study	136 BC to present	Climate, fire variability
Jemez Mountains	Touchan et al. 1995	Tree-ring fire scar study	AD 1600 to present	Grazing and fire
Jemez Mountains	Vierra et al. 1996	Archaeology	11,000 BP – AD 1500	—
Mesa Verde	Burns 1983	Crop yield simulation	AD 652–1968	Climate variability
Mesa Verde	Floyd 2003	Fire history	AD 1840 to present	Climate variability, grazing, fire suppression
Mesa Verde	Petersen 1994	Pollen	9800 BP to present	Climate variability
Mesa Verde (La Plata Mtns)	Petersen & Mehringer 1976	Pollen	9800 BP to present	Climate variability
Mesa Verde	Woodhead 1946	Range study	AD 1935–1946	Removal of cattle
Mesa Verde	Wycoff 1977	Pollen	AD 1300 to present	Deforestation (human)
Mesa Verde	Wilshusen & Towner 1999	Archaeology	Unknown	—
Northern Arizona	Blinn et al. 1994	Lake core diatoms, pollen, sediment chemistry	11,000 BP to present	Climate variability
Northern Arizona	Cole 1990	Packrat middens	14,000 BP to present	Climate, vegetation change
Northern Arizona	Salzer 2000a	Tree-ring study	AD 581 to present	Climate variability
Northern Arizona	Weng & Jackson 1999	Lake core pollen, charcoal, and sediment	14,000 BP to present	Climate, fire variability
Wupatki, Arizona	Cinnamon 1988	Packrat middens	12,000 BP to present	Climate, vegetation change

than 1 percent in fire-suppressed doghair thickets nearby (Ecological Restoration Institute 2002). Although the extent of change in heterogeneous ponderosa pine forests varies with site history, such changed conditions are now presumed to have affected millions of acres of ponderosa pine forests in the Southwest (U.S. General Accounting Office 1999).

The third vegetation type, pinyon-juniper woodlands, is much more heterogeneous than the other two types. These woodlands include at least 32 pinyon-dominated habitat types and 23 juniper-dominated habitat types, which range from ecotones with ponderosa pine forests, through closed pygmy woodlands, to juniper-dominated savannas and recently invaded grasslands (Mac 1998). Unfortunately, the fire histories and stand ages of most of these pinyon-juniper woodland habitat types remain poorly known, as substantial scientific uncertainty persists regarding the reliability and representativeness of fire scar evidence on either pinyons or junipers (Baker and Shinneman 2004; Floyd et al. 2004). National fire plans and assessments of the health of pinyon-juniper woodlands that have presumed frequent low-severity fires to be the norm have recently been found to be based on erroneous assumptions (Baker and Shinneman 2004; Floyd et al. 2004). In fact, two recent systematic reviews suggest that spreading, low-severity surface fires were not common in the historic record compared to infrequent, high-severity fires (Baker and Shinneman 2004; Floyd et al. 2004).

Some of the highest levels of species richness in the western United States have been found in landscapes dominated by pinyon-juniper woodlands, such as at Bandelier National Monument in the Jemez Mountains, Mesa Verde National Park, Canyon de Chelly National Monument, and Gray's Ranch in the Animas Mountains. It is nevertheless difficult to generalize about pinyon-juniper woodlands in terms of ecosystem health, biodiversity, or resilience, because this catch-all term includes ancient woodlands as well as recently invaded grasslands

and scrublands. Drought-aggravated bark beetle infestations and high-severity fires can completely destroy stands of pinyons and junipers, so that the oldest cohorts are often less than 200 years of age. Large wildfires in the pinyon-juniper woodland tend to kill all trees, seedlings, and propagules, leaving recovery dependent on reintroduction of seed by wildlife (Floyd et al. 2003b; Floyd-Hanna and Romme 1995). It may then take an additional 25 to 30 years for pinyon and juniper trees to reach reproductive maturity (Omi and Emrick 1980), with variable time periods in between reproductive, or "mast" years (Romme et al. 2003b). It may take about 300 years for a mature pinyon-juniper woodland to return to these earlier conditions following a catastrophic fire (Omi and Emrick 1980). Fire in this woodland community tends to leave a relatively barren, highly disturbed habitat that is vulnerable to invasion by exotic plant species such as musk thistle (*Carduus nutans*) and others (Floyd et al. 2001; Romme et al. 2003a). The fluctuating dynamics of age classes among pinyon and juniper vegetation often leave land surfaces vulnerable to erosion, especially as many of the open woodlands occur on sand, sandstone, and shales. In one survey of New Mexico woodlands, soil erosion was evident on 78 percent of some 1000 pinyon-juniper plots sampled (Van Hooser et al. 1993); loss of native bunch grasses (often a result of inappropriate livestock management) also contributes greatly to erosion (Davenport et al. 1998).

Although mixed conifer and ponderosa pine forests have been logged in many places, the recent historic changes in pinyon-juniper woodlands appear to be more related to livestock grazing intensities and shrub control programs. At least 600,000 acres of pinyon-juniper woodlands were treated mechanically in the 1950s and 1960s in Arizona alone (Mac 1998). In addition, herbicides were applied to kill junipers and pinyons on hundreds of thousands of acres through the 1980s as a means to improve forage conditions, but these treatments were questionable both economically and ecologically (www.mpcer.nau.edu/pj/blm_project _new/pji_main_tmpl.htm). Fuelwood cutting has occurred for centuries adjacent to pueblos, rancherias, towns, and villages in these woodlands, though cutting declined locally after the 1970s (Mac 1998); this may change as beetle-killed trees become available for cutting without the need for permits in some national forest districts.

In summary, at least 45 million acres of pine-dominated ecosystems occur in the western United States. There are considerable differences among the many habitat types of ponderosa- and pinyon-dominated vegetation on the Colorado Plateau, and caution should be taken in extrapolating from one habitat type to others. In particular, there is less certainty that most pinyon-juniper woodland habitat types have undergone dramatic changes in the frequency of high-severity fires passing through them over the last century and a half, relative to the dramatic changes documented in many mixed conifer and ponderosa forests.

These woody vegetation types harbor high levels of biodiversity. The Colorado Plateau as a whole—an ecoregion dominated by these three vegetation types—is ranked among the top 4 of 109 ecoregions in North America for species richness among several taxonomic groups, and is ranked first in levels of endemism (Nabhan et al. 2002a; Ricketts et al. 1999a, 1999b). Because so many of the ecoregion's endemic plants are edaphic endemics, it is clear that the diverse exposed geological strata of the Colorado Plateau have played a key role in the adaptive radiation of several important floristic groups. This geographic heterogeneity—and its influence on endemic, rare, threatened, and endangered species—underscores the need for resource managers to cautiously base their restoration and management strategies on localized or site-specific scientific studies of reference conditions and processes. We can no longer assume that one-size-fits-all, or single-point-in-time formulae will work for all pine-dominated ecosystems (Allen et al. 2002; Baker and Shinneman 2004; Romme et al. 2003b, 2003c; Savage 2003).

Fire History

Our review of fire history studies on the Colorado Plateau clearly demonstrates asynchrony in the dates when dramatic declines in fire frequencies occurred in southwestern ponderosa pine forests (Table 2). Using fire scar data from various sites first compiled by Savage (1989) and Savage and Swetnam (1990) and then expanded by our own literature search, it has become clear that the dates of historic fire frequency declines vary from 1750 to 1898. Two very different sites in Zion National Park in southwestern Utah differed by 12 years in their dates of decline, whereas two experimental forests in north-central Arizona differed by 22 years. As we shall see in the following discussion, there are many factors contributing to these differences in timing, including the timing and intensity of livestock grazing and logging, whether sheep or cattle were the predominant stock type, whether livestock were first introduced by sedentary or transhumant herders, and characteristics of the substrate, precipitation, and aspect of individual sites.

Pyne (2001) has offered new insights regarding the establishment of national fire suppression policy in 1912 following the inability of the Forest Service to control the western wildfires during a period known as the "Big Blowout." This policy's enactment varied considerably from place to place, however, with a time lag occurring before its imperfect adoption on tribal lands and in many remote areas where fire detection abilities and fire-fighting technology lagged behind. In any case, fire suppression due to overgrazing often predated formal fire suppression policy (see Table 2). However, after that policy was formally implemented it was difficult to convince local foresters of the value of managing for frequent low intensity burns.

Livestock History

Domestic livestock grazing in the West has long been a contentious issue, with extreme positions taken on both sides; however, recent initiatives to find workable solutions have provided some of the greatest

Table 2. Asynchrony in fire frequency decline in southwestern ponderosa pine forests.

Location	Date of Fire Decline
Chuska Mountains (northeast Arizona and northwest New Mexico)	1830
Horse Pasture Plateau (Zion National Park, southwest Utah)	1881
Church Mesa (Zion National Park, southwest Utah)	1893
Fort Valley Experimental Forest (north-central Arizona)	1876
Long Valley Experimental Forest (north-central Arizona)	1898
Gila Wilderness (Gila National Forest, southwest New Mexico)	1892
Prescott National Forest (north-central Arizona)	1874
Jemez Mountains (north-central New Mexico)	1750

success stories of collaboration among stakeholders in the region (e.g., see www.quivera coalition.org and www.diablotrust.org). These recent collaborative efforts have introduced possibilities for alternative management of livestock that will be less destructive to riparian areas and to native biodiversity. New management strategies such as grass banks and rotational grazing help preserve native grasses and ecosystems while also preserving the ranching way of life and regional food production capabilities.

The fact remains, however, that the introduction of domestic livestock brought about the first big impact of Europeans on the western landscape (aside from human depopulation resulting from the introduction of disease). It represents the most widespread influence on native ecosystems in western North America (Figure 1; Fleischner 1994; Wagner 1978). The influence of overgrazing on fire suppression and the disruption of ecosystem processes has been profound; there are no sites left in the West (other than isolated mesa tops) that have escaped livestock grazing in some form or other (Floyd et al. 2003c). There are difficulties in demonstrating the effects of grazing because the severity of these effects can vary so greatly depending on local conditions, type of stock and stocking rates, annual climatic variability, and different management practices (such as rotational grazing). In addition, there are few "control" areas because so few lands in the West have been free of

grazing influences. We lack benchmarks for pre-grazing conditions. Nevertheless, it is likely that the effects of livestock on a previously ungrazed landscape are dramatic, and recovery may not happen for decades or centuries—if at all—representing a threshold of change after which entirely new ecosystem dynamics have been put in place (Fleischner 1994; Westoby et al. 1989). In an extensive review of grazing impacts in the West, Fleischner (1994) included many impacts that persist over long periods of time, such as the effects of trampling and soil compaction, selective grazing on native bunch grasses and their inability to recover, and introduction of exotic weed species. Other lasting side effects include the extermination of predators by ranchers (Binkley et al., in review) and the alteration of hydrologic regimes to serve livestock (well digging, spring alteration, tank construction; Belsky et al. 1999; Woodhead 1946).

Comparisons of currently grazed, long-protected, and more recently protected sites in a 33,606 acres grazing exclosure at Chaco Canyon National Historic Park, in sagebrush scrub and grasslands, have confirmed what many ecologists have suspected: that plant species richness is typically higher on long-protected sites than on those under some grazing management regimes, regardless of soils, potential vegetation, or other management factors (Floyd et al. 2003c). Unfortunately, none of the study sites investigated in the Chaco Canyon area occurred in

Figure 1. Cattle round-up in Moencopi Wash near Tuba City, Arizona, 1941. Photo courtesy of Northern Arizona University.

pinyon-juniper woodlands or ponderosa pine forests. However, at Capitol Reef in Utah, Cole et al. (1997) determined that the most severe vegetation changes of the last 5400 years occurred as a result of the last century and a half of livestock overgrazing. Nevertheless, current experiments with rotational grazing do not necessarily result in de facto loss of biodiversity; the intensity and seasonality of grazing can result in widely varying effects that may not be the same for primary productivity, overall species richness, and the population viability of the rarest forage species on a site (Loeser et al. 2001).

Plant and soil characteristics across much of the Colorado Plateau indicate that these habitats probably evolved with low levels of soil surface disturbance by ungulates. Dominant native bunch grasses—many of them cool-season grasses—lack adaptations to grazing, such as tillering, secondary compounds, or high silica content (Belnap 2003; Parmenter and Van Devender 1995). In habitats that evolve with high levels of disturbance or grazing by large ungulates (such as bison), the dominant native grasses tend to be rhizomatous, and respond well to grazing and disturbance through a process called overcompensation (Bartolome 1993; Stebbins 1981). Loss of native bunch grasses on the Colorado Plateau has probably been an important contributor to erosion, a problem that has been well documented in the Jemez Mountains of New Mexico (Davenport et al. 1998). These native bunch grass ecosystems have been heavily impacted across the West by livestock (Bohrer 1975; Orodho et al. 1990). Historic accounts provide telling descriptions of the change in vegetation immediately after the introduction of livestock. In 1880, Jacob Hamblin wrote that all of the formerly fertile places on the Kaibab Plateau, which he had initially seen in 1870, were now occupied by white settlers, and that the grasses that had been so abundant and important to the Paiute were "all eat out by stock" (Fowler and Fowler 1971). Bohrer (1975) has also documented a loss or decreased abundance of cool-season grasses across the Southwest, especially in the context of ethnobotany. Many of the cool-season grasses represented important food resources for Native Americans before the introduction of livestock.

Biological soil crusts—delicate symbioses of cyanobacteria, lichens, and mosses—are well developed and widespread throughout the Colorado Plateau, where many ecosystems are dependent on the nitrogen provided by them (Belnap 1995; Evans and Belnap 1999). Biological soil crusts also provide important protection from wind and water erosion, surface albedo and temperature regulation, and carbon fixation (Belnap 2003). Soils that have not evolved with high levels of disturbance depend more heavily on soil surface integrity for natural ecosystem function than in other regions, and are negatively affected by disturbance (Belnap 2003). Recovery following trampling and compaction is very slow in these soils. Belnap (2003) has estimated that recovery rates

for biological soil crusts in southeastern Utah are approximately 15 years for cyanobacterial mass, 45–85 years for lichen cover, and 250 years for moss cover (mosses and lichens contribute up to 40% of undisturbed crusts). Even after 30 years without grazing, soil and plant nitrogen and nitrogenase activity levels in disturbed plots were significantly lower than in comparable ungrazed plots (Evans and Belnap 1999). Thus, recovery from grazing in an area formerly fixed by biological soil crusts for protection from erosion, soil moisture conservation, and nitrogen fixation could take up to 15 years to even begin recovery, and 250 years or more to recover the ecosystem function lost as a result of disturbance.

Early Anglo farmers and ranchers coming from the eastern United States where vegetation was adapted to grazing by bison (and thus fundamentally different in ability to "carry" domestic livestock) marveled at the lush vegetation and excellent forage when they first arrived. In short order they overstocked the range. Sheep stocking rates increased quickly and early; before 1846 there were an estimated 3 million sheep grazing on the eastern half of the Colorado Plateau (Baxter 1987). Cattle stocking rates were somewhat slower to establish, but in Arizona and New Mexico the rate increased from about 172,000 head in 1880 to more than 1.5 million in 1890 (Baker et al. 1988). In spite of early and irrefutable evidence of overgrazing's impacts, stocking rates continued to rise dramatically across Arizona and New Mexico to a peak around 1920, after which rates began a steady decrease, which has continued to the present (Dahms and Geils 1997).

The intensity and duration of grazing has varied greatly across the Colorado Plateau, as the histories of our four study sites indicate. For example, in the Jemez Mountains, grazing—predominantly by sheep—began early, was intense, and continued unbroken over hundreds of years. At Mesa Verde, grazing was introduced late and lasted only about 60 years before the park's establishment curtailed it, with a predominance of cattle and few sheep. In the Chuska Mountains and Canyon de Chelly, grazing of sheep began relatively early, has been very intense, and continues to this day. On the eastern flanks of the San Francisco Volcanic Field near Sunset Crater, sheep were established among the Navajo in relatively low densities, but by the 1880s Mormon ranchers brought in large numbers of cattle. The establishment of two national monuments did not fully eliminate livestock, which still persist in low numbers today. The relative grazing and browsing pressures by various native herbivores versus domestic livestock vary greatly at each site.

Introduction of Sheep into the West

An environmental history of the livestock trade in New Mexico provides a chronology for the introduction and spread of sheep and cattle in the Colorado Plateau region (Baxter 1987). Contrary to often-repeated assertions (Christman et al. 1997) that sheep, cattle, and horses were permanently established on the Colorado Plateau by Coronado and Arellano (1540–1541), or by Gallegos and Rodriguez (1578), it is more likely that the first lasting herds resulted from Onate's 1598 delivery to the upper Rio Grande of 3000 Churro sheep, 1000 Criollo cattle, 1000 goats, and 150 horses (Baxter 1987). The easternmost Navajos first acquired sheep around 1704, by raiding the Tewa Pueblos on the Rio Grande and its upper tributaries. This is quite different from the popular notion among contemporary Navajo Churro sheep fanciers that "the Navajo weaving tradition was born when the Spanish brought their Churro-type sheep to New Spain in the 1500s" (Christman et al. 1997:98), or that Churro sheep were selected and developed as a distinctive breed by the Navajo soon after 1540. We must also remember that Churro ancestors not only inhabited semi-arid portions of the Iberian Peninsula, but also evolved in the arid steppe regions of the Middle East and North Africa, essentially pre-adapting them to conditions on the Colorado Plateau.

In fact, sheep were first introduced in significant numbers to the eastern reaches of the Colorado Plateau landscapes somewhat after 1600, and were first used by the Navajo

around 1704 (Baxter 1987). Although undeniably a very important impact upon the land later in time, Navajo sheep herding probably had limited initial impact on the land anywhere until after 1818, when their raiding of Hispanic and Pueblo communities' stock accelerated. Throughout this early period of livestock adoption, sheep numbers were three to eight times higher than those of cattle (Baxter 1987). The peak years of sheep production and trade on the eastern half of the Colorado Plateau were 1820 to 1846, when upwards of 3 million head were recorded in the region (Baxter 1987). Sheep and cattle were not driven west of the Chuskas until Mormons brought them into Utah in 1847. Beale brought livestock through the Flagstaff area in 1857. Sheep ranching was established in the headwaters of the Little Colorado River around 1866 by Hispanic families (Baxter 1987), and Mormons initiated cattle ranching in the Virgin Valley of Utah in the 1860s (Madany and West 1983). Livestock was introduced at Mesa Verde and the Kaibab Plateau around 1870, with intense grazing pressure evident by the late 1880s. The most abusive overstocking of the New Mexico range, of both cattle and sheep, occurred in the 1880s through the early 1930s (Figure 2).

The purpose of this summary of the impacts of livestock is to offer insight to managers who must deal with the cumulative effects of grazing. Great variation in the timing of the introduction of livestock, the dominance of different types of stock, management practices, and the timing of stock exclusions have resulted in widely varying impacts on landscapes.

Logging and Shrub Removal

Selective logging and fuelwood cutting undoubtedly impacted prehistoric forests across the Colorado Plateau. In a study of prehistoric Puebloan fuelwood harvests, Kohler and Matthews (1988) demonstrated localized deforestation evidenced through changes in fuelwood harvesting preferences. There is also strong evidence for the selective harvest of large ponderosas, spruces, and firs from the Chuska Mountains during

Figure 2. Sheep crossing the Rio Grande River on the Buckman Bridge, 1922. Photo courtesy of Bandelier National Monument (National Park Service, BAND Cat. 14336, image 1608A).

large-scale construction episodes in Chaco Canyon, Canyon de Chelly, and Canyon del Muerto between AD 1030 and 1120. An estimated 200,000 trees were harvested just for prehistoric pueblo construction at Chaco Canyon (Betancourt et al. 1986); recent analyses have confirmed that these 200,000 trees came from several mountain ranges, and were derived from selective thinning of even-age cohorts rather than clearcutting (English et al. 2001).

These prehistoric logging activities had little impact on the landscape compared to the large-scale logging that began in the 1880s and increased rapidly through the twentieth century (Covington 2003; Dahms and Geils 1997; Friederici 2003b). In 1883, for example, the first sawmill in Flagstaff, Arizona produced 20 million board feet of lumber, mostly for export to the eastern United States (Figure 3; Friederici 2003b). From that time on logging essentially consisted of

clearcutting (Figure 4), until regulations were enacted in 1989 that were supposed to protect "seed trees" (Friederici 2003b). Timber harvests on national forest lands, tracked since 1908, gradually increased through the 1980s (Dahms and Geils 1997), but after 1990 harvests decreased substantially as a result of environmental concerns, tighter regulations, and a relative paucity of large trees (Covington 2003; Dahms and Geils 1997). In fact, the harvest of 46 million board feet from Forest Service lands in 1996 was primarily for fuelwood (Dahms and Geils 1997). By this time, old-growth ponderosa forests in Colorado and New Mexico were reduced to less than 5 percent of their former range, a situation reflected elsewhere in the Southwest as well (Figure 5; Covington 2003).

During these intensive logging activities, recruitment of new ponderosas beginning around 1912 led to dense doghair thickets of

Figure 3. Arizona Lumber and Timber Company sawmill, 1899. Photo courtesy of Northern Arizona University.

thirsty young trees, which consumed larger amounts of water than the more open stands of mature trees that typify old-growth ponderosa forests. The combination of these dense stands of young trees and the slash piles left by logging activities increased the fuel load dramatically, thus greatly heightening the intensity of natural wildfires (Covington 2003).

Simply understanding these general patterns of cumulative effects in ponderosa forests does not in itself offer a clear formula for making management decisions. There is great variability in the impacts of logging across the Colorado Plateau. Within our four study areas, we have documented little to no historic logging at Mesa Verde and in the Wupatki and Sunset Crater areas, but heavy

and destructive logging in portions of the Chuska and Jemez Mountains.

Prehistoric fuelwood cutting also had an impact on localized pinyon-juniper habitats; in historic times, however, the greatest influences on the distribution of pinyon-juniper woodlands have been from domestic livestock, and subsequent pinyon-juniper expansion followed by eradication efforts. Following the introduction of livestock into western grasslands and pygmy woodlands, understory herbaceous cover decreased dramatically, the frequency of low-intensity surface fires declined because of this lack of fine fuels on the ground surface (Romme et al. 2003a), and some woodlands became more dense and began "invading" grasslands (Allen 1998). These changes in the

Figure 4. Logging camp south of Flagstaff, 1904. Photo courtesy of Northern Arizona University.

distribution of pinyon and juniper began to attract the attention of range managers, as their encroachment into grasslands became an increasing problem for ranchers, stimulating eradication measures that have become increasingly sophisticated—although perhaps not more effective. The Merriam-Powell Center for Environmental Research (MPCER) at Northern Arizona University is involved in an ongoing study of eradication efforts and their effects over the past 57 years on Bureau of Land Management (BLM) lands on the Colorado Plateau (www. mpcer.nau.edu/pj/). Pinyon-juniper woodlands constitute the largest vegetation type managed by the BLM on the Colorado Plateau. The MPCER study is intended to provide information regarding the extent and effectiveness of past pinyon-juniper removal efforts. Preliminary results indicate that there have been two well-defined

periods of pinyon-juniper removal on BLM lands: the "Chaining Era" from 1950 to 1979, characterized by chaining and/or bulldozing, and the "Diversified Methods Era" from 1980 to the present, when prescribed burning dominated, but other methods such as hydroaxe, rollerchopping, hand thinning, and chemical treatments were also widely used. The trend in types of treatment methods used and the number and size of treated areas is toward more use of prescribed burns and rollerchopping in the northern and northeastern portions of the Colorado Plateau, and a greater emphasis on chemical treatment on the western plateau. Perhaps of greater concern is that both the sizes of treatments and the numbers of treatments have increased dramatically since 2000 (Rundall et al. 2003). Although the MPCER project is designed to provide background information regarding the history of

Figure 5. Logging camp south of Flagstaff, 1904. Photo courtesy of Northern Arizona University.

these treatments, information regarding the effect of the treatments is lagging behind the rapid expansion of their use.

Indigenous Fire Management

Fire, whether natural or intentional, produces a mosaic landscape, and stimulates the growth of some plants (Table 3). Some of the plants that require fire or smoke for germination, such as wild tobacco, are or were important to native people; others, such as the straight shoots of some shrubs that are so valuable for basketry, may not require fire for germination, but become more useful after fire. Because remains of many of these plants have been found in archaeological contexts, it is apparent that fire played an important role in ensuring the availability of these resources.

One of the most hotly debated factors in vegetation change in the Americas is the relative importance of indigenous fire management in prehistoric and historic times (Dahms and Geils 1997; Nabhan 1998). It is not that researchers, resource managers, or conservation activists doubt that indigenous

communities on the Colorado Plateau and elsewhere intentionally set fires to manage a variety of natural and cultural landscapes (Table 4); it is simply the magnitude of these fires and their effect on ecosystems that remains debated.

The current controversies surround three related questions: (1) Were Colorado Plateau landscapes so strongly shaped by natural processes such as lightning-ignited fires that human-ignited fires (Allen 2002a) were essentially masked? (2) Was the use of fire by prehistoric and historic cultures of the Colorado Plateau as intensive and extensive as that recorded in other regions such as in the California Sierras, the Great Basin, and the Plains (Alcoze 2003; Anderson 1993)? (3) Did the significant cultural diversity within the region create variation in the degree to which Puebloan farmers, Athapaskan herder-hunters, and Rancherian hunter-gatherers created culturally managed landscapes through their use of fire, fields, and other means (Nabhan et al. 2002a)?

Given its richness in long-term paleoecological, dendrochronological, archaeological,

Table 3. Fire-stimulated plants.

	Scientific Name	Common Name	Regeneration After Fire or Disturbance	Comments
SHRUBS				
Agavaceae	*Yucca baccata*	Banana yucaa	Resprouts from rhizomes or underground stem buds.	—
Ancardiaceae	*Rhus aromatica*	Fragrant skunkbush	Little information.	Resprouts in Great Plains.
Anacardiaceae	*Rhus trilobata*	Skunkbush	Resprouts from root crown or rhizomes.	May be able to delay sprouting for up to 1 year, an adaptation favorable in harsh climates; may reproduce from seed bank after fire.
Asteraceae	*Chrysothmanus depressus*	Rabbitbrush	No information.	Other C. spp. resprout from roots or stems (*C. viscidiflorus* and *C. nauseosus*).
Caprifoliaceae	*Symphoricarpos oreophilus*	Snowberry	Limited resprouting from root crown.	—
Fagaceae	*Quercus gambelii*	Gambel oak	Resprouts from adventitious buds on rhizomes and lignotubers, and from stem sprouts.	Sprouts documented within 10 days of fire (Tiedmann et al. 1987).
Fagaceae	*Quercus turbinella*	Canyon live oak	Resprouts vigorously from root crowns and rhizomes; seedling establishment after fires.	—
Grossulariaceae	*Ribes aureum*	Golden currant	Resprouts.	—
Rhamnaceae	*Rhamnus smithii*	Buckthorn	No information.	Other species of *Rhamnus* (*R. californica* and *R. pushiana*) sprout from root crowns after top is fire-killed.
Rosaceae	*Amelanchier utahensis*	Utah serviceberry	Resprouts from root crown.	Slow recovery.
Rosaceae	*Cercocarpus montanus*	Mountain mahogany	Resprouts from root crown.	Burns less readily than other shrubs.
Rosaceae	*Prunus virginana*	Chokecherry	Resprouts rapidly and prolifically from root crowns and rhizomes.	Seed germination improves with heat treatment (Sampson 1944).
Rosaceae	*Purshia tridentata*	Bitterbrush	Resprouts from root crown, burns on or below ground lignotuber, or meristematic tissue beneath bark.	—
Solanaceae	*Lycium* spp.	Wolfberry	Resprouts from root crown.	—

Table 3 (continued)

	Scientific Name	Common Name	Regeneration After Fire or Disturbance	Comments
HERBS, FORBS, AND GRASSES				
Asteraceae	*Helianthus annuus*	Sunflower	No information for *H. annuus*, but at least one species of sunflower, *H. maximiliani*, root sprouts after fire.	—
Boraginaceae	*Hackelia gracilenta*	Stickseed	Rapid seedling release after fire at Mesa Verde National Park (Adams 2002).	—
Brassicaceae	*Schoencrambe linifolia*	Perennial mustard	Resprouts from caudex and underground rhizomes after disturbance.	—
Chenopodiaceae	*Chenopodium fremontii*, *C. berlandieri*, *C. leptophyllum*	Goosefoot	Rapid seedling release after fire at Mesa Verde National Park (Adams 2002).	—
Fabaceae	*Lupinus caudatus*	—	Rapid seedling release after fire at Mesa Verde National Park (Adams 2002).	—
Malvaceae	*Sphaeralcea coccinea*	Globemallow	Top-killed, resprouts from rhizomes; vigorous growth after fire.	—
Liliaceae	*Calochortus nuttallii*	Sego lily	Rapid seedling release after fire at Mesa Verde National Park (Adams 2002).	
Poaceae	*Achnatherum* (*Oryzopsis*) *hymenoides*	Indian ricegrass	Resprouts from tillers and reseeds from plants outside fire boundary.	On burned CO pinyon (*Pinus edulis*)–Utah juniper sites at Mesa Verde, Indian ricegrass & other perennial grasses dominated the site by the 4th year after fire (Erdman 1970).
Poaceae	*Poa fendleriana*	—	Probably sprouts from rhizomes, as *P. compressa* (Canada bluegrass) and *P. pratensis* (Kentucky bluegrass) do.	Postfire plant vigor & density greatly affected by phenological stage at time of burning.
Polygonaceae	*Eriogonum* spp.	Buckwheat	Rapid seedling release after fire at Mesa Verde National Park (Adams 2002).	—
Polygonaceae	*Polygonum douglasii*, *P. sawatchensis*	Knotweed, pinyon knotweed	Usually reproduces from seed after fire.	Reproduction increased after severe fire; peaks and then declines following fire.
Scrophulariaceae	*Penstemon caudataus*	Lupine	Plant resprouts from caudex after disturbance.	—
Solanaceae	*Nicotiana attenuata*	Coyote tobacco	Seeds germinate from smoke (Baldwin et al. 1994).	Peaks then declines after fire.

Source: USDA Fire Effects Information http://www.fs.fed.us/database/feis/plants/

Table 4. Use of fire among three broad cultural subsistence groups on the Colorado Plateau.

	Puebloan Farmers	Nomadic Hunters/ Herders	Rancherian Hunter- Gatherers
Promote hunting	X	X	X
Promote basketry materials growth	X	X	X
Fireproofing areas of dwelling	X	–	–
Promote plant foods	X	X	X
Clear fields	X	X	–
Collect insects	–	–	X
Manage pests	X	–	–
Fell trees	–	–	X
Signal allies	–	X	X
Starve enemies or their horses	–	X	–

Sources: Alcoze 2003; Allen 2002a; Bohrer 1983; Kaib 1998.

and ethnobotanical information, the Colorado Plateau provides a unique opportunity to assess the relative importance of both natural and cultural factors in the ecological history of fire (Alcoze 2003; Alcoze and Hurteau 2001; Allen 2002a). Thus, it may be possible to test the null hypothesis that fire frequencies, intensities, and areal extents were no different during particular prehistoric and historic periods of cultural occupation of certain landscapes of the Colorado Plateau than they were before or after.

To reject this null hypothesis, evidence of cultural fire influences cannot be masked by lightning fires in season or frequency. And yet, as the many sources cited by Allen (2002a) suggest, a key feature of the Colorado Plateau ecoregion is the plentiful source of natural fire ignitions through extremely high levels of lightning activity. In a study area centered on the Jemez Mountains covering roughly 1.8 million acres, more than 165,000 cloud-to-ground lightning strikes were recorded in just one decade. With the annual number of lightning strikes ranging from 9400 to 23,400 in this study area, it is difficult to imagine how human-ignited fires could be discerned in the tree-ring record of fire scars, unless they were out of season. As Bahre (1991:128) concluded from the upland southwestern landscapes south of the Colorado Plateau, "given the high frequencies of

lightning-caused fires reliably documented by modern data, the relative importance of fires set by Indians is probably moot."

Allen (2002a) generally agrees with Bahre (1991), interpreting current data to conclude that the strong consistencies between the multicentury prehistoric fire-scar record and observations of modern lightning-caused wildfire activity suggest that fire regimes before 1900 can be mostly or even wholly attributed to natural factors.

The alternative view that indigenous burning was a major influence on the Colorado Plateau harkens back to observations made by John Wesley Powell before 1890, although Dobyns (1981) has argued that Spanish colonists made similar observations more than two centuries earlier. Powell's (1890) "Paiute forestry" hypothesis suggested that a fire management system was skillfully used not just by the Paiutes but by other Southwest tribes as well:

Before the white man came, the natives systematically burned over the forest lands with each recurrent year as one of their great hunting economies. By this process, little destruction of timber was accomplished; but protected by civilized men, forests are rapidly disappearing. The needles, cones and brush, together with the leaves of grasses and shrubs below, accumulate when not burned annually. New deposits are made from year to year, until the ground is covered with a thick mantle of flammable material. (Powell cited in Alcoze 2003: 51)

How Powell—from his brief visits with the Southern Paiute tribe—could have documented that burning was intentionally done annually on either the same or neighboring plots, remains a mystery. Indeed, although there is no doubt that the Southern Paiutes have extensive traditional ecological knowledge of the effects of burning on plants and wildlife on the Colorado Plateau, more recent interviews with Paiute elders have not revealed much more detail to outsiders, only some minor differences compared to Powell's generalization (Alcoze 2003). Paiute elder Pikyavit, when interviewed at Pipe Springs National Monument where he has served as a ranger, has revealed this much: "Southern Paiute people burned pine groves on a four-year cycle to maintain open woodland structure, reduce understory litter, increase nut productivity, and reduce insect and disease damage" (Pikyavit, paraphrased in Alcoze 2003). Alcoze (2003) has suggested that Paiute fires set in pinyon-juniper woodlands sometimes reached into adjacent ponderosa pine patches, but this conclusion is based on academic studies rather than on verbatim quotes from any Southern Paiute elders that verify the burning of ponderosa forests.

Indeed, as Allen (2002a) has conceded after an exhaustive literature review of some primary and many secondary sources, "Primary evidence for landscape-scale burning for hunting purposes is nearly nonexistent in the Southwest, and supporting rationales are weak ... Evidence of landscape-scale fire use by aboriginal people in the Southwest (for *any* purpose) is scanty to nonexistent, and most assertions of aboriginal burning are based on anecdotal accounts or sources subject to substantial historical bias."

It is Allen's as well as our impression that few ethnographers have documented the particularities of indigenous fire management on the Colorado Plateau to the extent that they have been documented elsewhere, although the examples provided below still suggest that culturally managed fire was a significant factor in some areas of the region during specific periods. Rather, the problem is that ethnohistorians like Dobyns (1981)

have been all too willing to make broad extrapolations from scant evidence, rather than spending their time interviewing additional elders for tangible details or undertaking fire scar studies to document the scale, intensity, and frequency of culturally ignited fires.

We agree with Allen that the degree of documentation of fire management is not as detailed in the Southwest as in other regions, but we are less willing to concede that the evidence is negligible. Indeed Kaib et al. (1996) and Kaib (1998) amply reviewed ethnohistoric documents of Southern Athapaskan (Apache and proto-Apache) fire use for a variety of purposes, not just for fire-drive hunting. Kaib (1998) also used fire scar data to document an upsurge in fires during the period of Apachean warfare (up through the 1880s) occurring outside the normal season, and differing from the previous fire return interval.

Monti (2003) has documented the continued traditional use of fire in the Chuska Mountains by Navajo farmers, who maintain a landscape mosaic of different-aged stands of vegetation regardless of official fire suppression policies. These Athapaskan traditions consistently show a frequency and variety in the use of fire to manage vegetation and wildlife that surpass known uses among Puebloan people. Additonally, oral histories and historic photographs from the Mesa Verde area of southwestern Colorado document the use of fire by Utes, especially in petran chaparral communities (M. Colyer, personal communication, 6 Oct 2003).

In short, there is considerable heterogeneity in the cultural influences on fire ecology on the Colorado Plateau. The dates when these aboriginal practices were formally suppressed also vary greatly, with the Chuska Mountains and White Mountains being perhaps the last areas where formal suppression came into effect.

Faunal (Avian) Diversity

Faunal diversity in ponderosa forests and pinyon-juniper woodlands is related to a variety of factors, some affected by management and others not. Table 5, derived from

Table 5. Hypothesized avian population changes from pre-1850 conditions, inferred from survey plots.

Species	Population Change in		Cause
	Ponderosa	Pinyon-Juniper	
Mourning dove	–	–	Decrease in grass seed, loss of open conditions
Flammulated owl*	–	nt	Fewer snags, holes
Mexican spotted owl	–	–	Fewer snags, loss of large trees
Purple martin*	–	nt	Fewer snags, old trees
Broad-tailed hummingbird	–	–	Decrease in flowers
Acorn woodpecker*	–	nt	Fewer oaks, snags
Hairy woodpecker*	nt	nt	—
Northern flicker*	nt	nt	—
Northern three-toed woodpecker*	–	nt	Decrease in old trees
Western wood-pewee	–	nt	Increase in doghair thickets, loss of open conditions
Gray flycatcher	nt	+	—
Ash-throated flycatcher*	nt	nt	Fewer old trees, snags, holes
Violet-green swallow*	–	–	Loss of open conditions, fewer cavities
Steller's jay	nt	nt	—
Mountain chickadee*	nt	nt	—
Juniper titmouse*	nt	nt	—
White-breasted nuthatch*	–	–	Fewer snags, oaks
Pygmy nuthatch*	nt	nt	—
Brown creeper	nt	nt	—
Rock wren	–/nt?	–/nt?	Loss of open conditions
Western bluebird*	–	–	Loss of open conditions
Mountain bluebird	–	–	Loss of open conditions, loss of snags for nesting
Townsend's solitaire	+	nt	Increase in forest density
Hermit thrush	+	nt	Increase in forest density
American robin	–	–	Decrease in understory
Plumbeous vireo	–	–	Increase in forest density
Virginia's warbler	–	–	Loss of understory, oaks
Yellow-rumped warbler	nt	nt	—
Grace's warbler	+	nt	—
Black-headed grosbeak	–	–	Decline in understory oaks and shrubs
Spotted towhee	–	–	Decline in shrubby understory
Chipping sparrow	–	–	Loss of open conditions
Lark sparrow	–	–	Decrease in grass seed and burns, loss of open conditions
Dark-eyed junco	nt	nt	—
Western tanager	–	–	Loss of large, old trees

*Cavity nester.
nt = no trend; + increase; – decrease.
Sources: Brawn and Balda 1988; San Miguel and Colyer 2003.

bird surveys, suggests that diversity can decline due to the proliferation of doghair thickets, the loss of understory plants (especially grasses), and the decline of herbaceous plant productivity. Other factors include the increase in exotic plant species (such as thistles), which prevail over natives; the loss of old trees, snags, and nest holes; a decline in pinyon nuts and juniper berries; and changes in fire frequencies and their effect on insect faunas. Although most populations of vertebrates respond to several of these factors, Brawn and Balda (1988) have attempted to demonstrate the effects of silviculture management on birds in the ponderosa pine forests above the Mogollon Rim in northern Arizona. We have reinterpreted some of their data, focusing not so much on the effects of logging and thinning as on the effects of fire suppression and other factors (i.e., exotic invasions). In addition, we have added complementary data from ancient pinyon-juniper woodlands around Mesa Verde in Colorado (San Miguel and Colyer 2003).

Predator Control

Predators exert a top-down regulatory influence on terrestrial ecosystems (Crooks and Soulé 1999; Henke and Bryant 1999; McLaren and Peterson 1994; Terborgh 2001; Terborgh and Winter 1980). The eradication of predators or other highly interactive species can have strong effects on the structure and stability (Mittlebach et al. 1995) as well as composition (Henke and Bryant 1999) of an ecosystem; it can leave a functional void that can trigger linked changes that lead to degraded or simplified ecosystems (Soulé et al. 2003). Recent research has shown that some of the impact of wolves and other predators occurs through their influences on prey populations. For instance, as a major predator of large ungulates, gray wolves may suppress prey levels or alter prey behavior to the extent that they affect vegetation patterns and productivity (Wilmers et al. 2003).

One of the first observations concerning top-down regulation of ecosystems by predators in the forestry literature was made by Aldo Leopold (1943) in reference to the Kaibab Plateau in northern Arizona. According to Leopold and the sources that he drew upon, the intensive hunting of cougars (*Felis concolor*), gray wolves (*Canis lupus*), and coyotes (*Canis latrans*) after the establishment of the Grand Canyon Game Preserve in 1906 led to an irruption in deer populations in the 1920s. The intensity of predator hunting on the Kaibab Plateau was such that between 1906 and 1931, government-hired hunters killed an estimated 781 cougars, 30 wolves, 4889 coyotes, and 554 bobcats (*Felis rufus*; Botkin 1990). The resulting deer population explosion led to overbrowsing, followed by a deer population crash, and then degraded habitat that persisted for decades. This hypothesis was later criticized by Burk (1973), Caughley (1970), and others because there were questions about deer population estimates, and other possible explanations were put forward for the reduction in the numbers of aspen shoots and trees of a certain cohort. Young aspen shoots comprise the majority of summer deer browse on the Kaibab Plateau (Bostick 1949). Recent examinations of aspen stand age structure on the Kaibab Plateau support the classic view of high deer populations in the 1920s, as proposed by Leopold (Binkley et al., in review; Moore and Huffman, in press). A similar situation on the Kaibab Plateau occurred again in the 1940s, when intensified predator control using poison baits coincided with a period of low predation by humans (hunting), which resulted in another deer irruption (Binkley et al., in review). Not only was aspen regeneration affected in these episodes, but Gambel oak and ponderosa pine were diminished as well (S. Buckley, personal communication 2004).

Although the amount of aspen habitat within our study areas varies greatly, the issue of predator control (and reintroduction) does apply to management of many natural areas on the Colorado Plateau. Overbrowsing and overgrazing by unchecked ungulate populations has serious implications for other vegetation types as well, including oaks, shrubs, understory grasses and forbs, and even fruit-bearing cactus species. Elk numbers in Arizona and New Mexico were higher in 1993 than in any period in recorded history (Irwin et al. 1993) during a period when predator numbers (other than coyotes) were low. The reintroduction of Mexican wolves to the southern Colorado Plateau may now serve as a test of the top-down regulation hypothesis.

Insect Outbreaks

Epidemic outbreaks of bark beetles over the past 4 years have decimated forests on the Colorado Plateau (De Gomez 2003). In Arizona and New Mexico, between 2000 and 2002, tree mortality from bark beetle infestations increased from 73,000 acres to 750 million acres of dead or dying trees (U.S. Forest Service 2002). By the end of 2003, an estimated 20 million ponderosa pines and 50 million pinyon pines had been killed by beetles in Arizona and New Mexico (Nijhuis 2004). Estimates of 80–90 percent tree mortality in some stands have been documented in the Tonto, Apache-Sitgreaves, and Prescott National Forests, and on the San Carlos Apache Reservation. Impacts of comparable severity have occurred in pinyon-juniper woodlands on the Colorado Plateau and eastward to the Pajarito Plateau in the Rio Grande watershed. A 2002 survey of 28 square miles of pinyon-juniper woodland southeast of Flagstaff, Arizona revealed 90 percent mortality of pinyon pines, or the loss of 700,000 mature trees (De Gomez 2003). Similar losses occurred over the same time period in Bandelier National Monument in New Mexico (C. Allen, personal communication 2003). Tens of thousands of acres of junipers and spruce were infested as well, although mortality was not as extensive as in ponderosa and pinyon pines.

The dramatic loss of Southwest pinyon and ponderosa pine trees has focused public awareness on bark beetles (De Gomez and Young 2002), but a host of other insects and pathogens can be devastating as well (Table 6). The cypress bark beetle is currently killing junipers and Arizona cypress (*Cupressus arizonica*) in the Southwest, and the spruce bark beetle is infesting spruce species at higher elevations. Root fungi (e.g. *Armillaria*

spp. and *Heterobasidion annosum*) and associated pests kill about 34 percent of trees greater than 5 inches in diameter (Wood 1983), dwarf mistletoes (*Arceuthobium* spp.) are specialized parasites of conifers, and white pine blister rust (*Cronartium ribicola*), an introduced fungus, is impacting white pines in the Sacramento Mountains of New Mexico (Conklin 2004; Dahms and Geils 1997). In 1934 and 1935 at Mesa Verde, 12,000 dead and dying pinyon pines were infested with black stain root disease (*Leptographia wagenerii*), a lethal fungal pathogen.

Insect infestations and tree diseases play an important ecological role by "thinning" forests and woodlands, and are an integral part of the food base for birds, predator insects, and other wildlife. There are historic records of significant outbreaks over the past century. Bark beetles killed large numbers of trees on the Colorado Plateau during the 1950s drought. That drought was not as severe at Mesa Verde, but there were severe outbreaks of pinyon pine beetle (*Ips*) infestations there in the 1930s and again in the 1970s (Floyd et al. 2003d; Omi and Emrick 1980). In southeastern New Mexico in the Sacramento Mountains, six outbreaks of roundheaded pine beetles have occurred since 1937 (Bennett et al. 1994). Generally small areas of forest were infected, but a larger outbreak in the 1970s, covering 150,000 acres, killed an estimated 400,000 second-growth ponderosa pines (Massey et al. 1977), which shifted the dominant tree species to Douglas fir and white fir. Large outbreaks of pine beetles also occurred on the Kaibab Plateau (Dahms and Geils 1997). Spruce beetle outbreaks occurred in New Mexico in the Jemez Mountains during the 1970s and in the Pecos Wilderness between 1982 and 1985, and in the White Mountains of Arizona in the 1980s (Bennett et al. 1994). Baker and Veblen (1990) used historic photographs and dendrochronology to reconstruct a history of spruce beetle outbreaks in western Colorado back to the 1900s; they concluded that spruce bark beetles have been a major disturbance agent, comparable to fire.

The mass tree mortality from insect outbreaks is not limited to the Southwest. Large

Table 6. Forest tree insects.

Primary Host Tree	Bark Beetles	Defoliating Insects	Shoot, Twig, Tip, & Sucking Insects
Ponderosa pine (*Pinus ponderosa*)	Roundheaded pine beetle (*Dendroctonus adjunctus*), western pine beetle (*D. brevicomis*), mountain pine beetle (*D. ponderosae*), red turpentine beetle (*D. vallens*), pine engraver (*Ips pini*), Arizona fivespine ips (*I. lecontei*)	Ponderosa needle miner (*Coleotechnites ponderosae*)	Pine needle scale (*Chionaspis pinifoliae*)
Pinyon pine (*Pinus edulis*)	Pinyon bark beetle (*Ips paraconfusus*)	Pinyon needle miner (*Coleotechnites edulicola*)	Pinyon tip moth (*Dioryctria albovitella*), pinyon needle scale (*Matsuococcus acalyptus*)
Juniper (*Juniperus* spp.)	Cypress bark beetle (*Phloeosinus cristatus*)	—	Juniper twig pruner (*Styloxus bicolor*)
Douglas fir (*Pseudotsuga menziesii*)	Douglas fir beetle (*Dendroctonus pseudotsugae*), fir engraver beetle (*Scolytus ventralis*)	Western spruce budworm (*Choristoneura occidentalis*), Douglas fir tussock moth (*Orgyia pseudotsuga*)	Pine needle scale (*Chionaspis pinifoliae*)
White fir (*Abies concolor*)	Fir engraver beetle (*Scolytus ventralis*)	Western spruce budworm (*Choristoneura occidentalis*), Douglas fir tussock moth (*Orgyia pseudotsuga*)	—
Southwestern white pine (*Pinus strobiformis*)	Roundheaded pine beetle (*Dendroctonus adjunctus*), mountain pine beetle (*D. ponderosae*)	—	Pine needle scale (*Chionaspis pinifoliae*)
Corkbark fir (*Abies lasiocarpa* var. *arizonica*)	—	Janet's false hemlock looper (*Nepytia janetae*)	—
Engelman spruce (*Picea engelmannii*)	Spruce beetle (*Dendroctonus rufipennis*), fir engraver beetle (*Scolytus ventralis*)	Western spruce budworm (*Choristoneura occidentalis*), Douglas fir tussock moth (*Orgyia pseudotsuga*), Janet's false hemlock looper (*Nepytia janetae*)	Spruce aphid (*Elatobium abietinum*), pine needle scale (*Chionaspis pinifoliae*), Cooley spruce gall adelgid (*Adelges cooleyi*)
Aspen (*Populus tremuloides*)	—	Large aspen tortrix (*Choristoneura conflictana*), tent caterpillar (*Malacosoma californicum*)	—

Adapted from the University of Arizona Cooperative Extension (2004), Arizona Health Web page (http://ag.arizona.edu/extension/fh/insects.html).

tracts of forest in the Pacific Northwest and Canada are dying from attacks by bark beetles, defoliating insects (McCullough et al. 1998; Nijhuis 2004), and other pathogens. The scale of the present die-off has researchers scrambling for causes and solutions. The epidemic is in part another legacy of fire suppression. The relationships between forests, insects, fires, and pathogens are dynamic and synergistic, affecting succession, nutrient cycling, and forest structure and composition. Fire suppression led to successive generations of homogeneous forest stands, decreasing landscape heterogeneity, and concentrating host trees in dense stands. Forest canopy closure and loss of patches, edges, and understory herbs and shrubs decreased important habitat for insect predators. Another factor exacerbating the current situation is drought, which has stressed and weakened over-crowded trees. But the proverbial last straw may be global warming (Lynch and Swetnam 2003; Nijhuis 2004), which translates into warmer winters and more frost-free days on the Colorado Plateau, resulting in decreased winter mortality of the insects and in some cases allowing a second generation in the same year. The combined effects of climate change, drought, and unusually dense forests has amplified the natural levels of insect outbreaks to crisis dimensions.

Because forest pests and diseases leave no direct record of their infestations in tree rings, the extent of prehistoric infestations is difficult or impossible to determine. Most episodes of reduced forest cover found in the prehistoric record have been determined on the basis of pollen or dendrochronological data, which do not provide evidence of the cause of deforestation. Thus some dramatic changes in forest and woodland density and composition which have been attributed to fire or tree cutting by humans could have been caused by bark beetle or other infestations. The most outrageous example of attributing woodland change to human deforestation on the Colorado Plateau is Jared Diamond's (1986) imperfect parable regarding the collapse of the Chaco Canyon Anasazi and prehistoric Puebloans,

as recounted in an awarding-winning book, *The Third Chimpanzee*, and in *The New York Times* and *Nature*. Diamond did no fieldwork in Chaco Canyon, drew upon but one source (Betancourt and Van Devender 1981) which did not come to the same conclusions, and ignored the possibility that bark beetle or other infestations were just as plausible a driver of woodland change as human uses. More recent studies indicate that the Chacoans did not clearcut surrounding forests and woodlands, but instead selectively cut single-aged cohorts of spruce trees for vigas (beams). These trees came from several mountain ranges within about 45 miles of Chaco Canyon (Betancourt et al. 1986; English et al. 2001). Until Diamond can refute the hypothesis that bark beetles, not woodcutting, led to the changes in frequencies of pine and juniper in packrat middens from the Chaco area, both natural and social scientists should be skeptical of his sweeping hypothesis that Chacoans destroyed their watershed environment and therefore caused the collapse of their own civilization. Unfortunately, his flawed parable has gained currency with other historians of Native American impacts on the Colorado Plateau (Krech 1999), despite the scant evidence upon which it was based.

There are no clear remedies for the current outbreaks. The use of controlled burns and natural fires to restore forest health can stress and weaken trees, making them more susceptible to bark beetle attacks (Ganz et al. 2003). New burns attract certain insects, such as *Melanophila* spp., which are wood borers with fire-homing infrared sensors on their legs that can guide these beetles to fires up to 3 miles distant (Evans 1973). The season when fires occur is also an important factor; in controlled burns in California, Ganz et al. (2003) found that tree mortality from bark beetle attacks following fires was lower in fall burns compared to spring and summer. Mechanical tree thinning can increase resistance to bark beetle attacks only if reduced competition among trees leads to greater water availability and thus increased vigor in the remaining trees (Preisler and Mitchell 1993). This strategy is unlikely to effectively reduce water stress during a drought as severe as the current one, and may in fact aggravate bark beetle infestations because beetles proliferate in logging slash; they can disperse to live trees in stands as much as 6 miles away (Preisler and Mitchell 1993). Thus, conventional commercial logging without slash removal can aggravate infestations, and may have minimal impact on reducing infestations as long as drought remains severe (Southwest Forest Alliance 2002).

Climate Change

The Colorado Plateau is characterized by exceptional biodiversity that is maintained in situ because the landscape is topographically complex, the climate is extremely variable, and human occupation is culturally mixed (Nabhan et al. 2002b). The phenomenon of climate is complex, dynamic, and seemingly impossible to predict, as it is created by multiple processes, each operating at different time scales. Astronomical and global mechanisms are the architects of climate, but there are also random events, such as meteors, volcanoes, and anthropogenic pollution, which can tip the scale to heat waves, cold periods, and unusual moisture conditions. Here we examine both paleo and modern records of climate from the perspective of how climate has influenced Southwest forests and woodlands.

The longest cycling process that changes global climate involves the amount of solar energy received at the earth's surface. Solar insolation changes in sync with the wobble in the tilt of the earth's axis and the orbital patterns of earth and sun at a scale of 100,000 years, with additional nested cycles of 41,000 and 19,000–23,000 years (Milankovitch 1941). Milankovitch's model predicts a rhythm of cold and warm climate changes; this system has been correlated to global records of glaciation and deglaciation. There are several other stochastic climate-changing processes as well as those occurring at high-frequency cycles, but the two most visible are volcanic eruptions and solar activity.

Reconstructions of the Holocene paleoclimate have been extrapolated from a variety of biological and physical fossils, such as pollen, charcoal, snails, insects, diatoms and other microorganisms, packrat middens, sediments and soils, and other more obscure proxy data (e.g. phytoliths and isotopes). Each discipline has strengths, biases, and a methodological myopia (see Table 7) that must be filtered and interpreted to discern a signal of climate trends (Egan and Howell 2001).

A coarse outline of the Southwest Holocene climate begins with the close of the last glacial cycle, some 12,000 years ago at the end of the Pleistocene. At this time, a generally synchronous major vegetation change is visible in the paleo records from packrat middens from the Great Basin and Southwest (Betancourt et al. 1990; Cole 1990) and from stratigraphic pollen records from lake sediments in the Chuska Mountains in northwest New Mexico (Wright et al. 1973) and the southern Colorado Plateau (Anderson 1993; Weng and Jackson 1999). These records show that at the Pleistocene to Holocene transition, ponderosa and pinyon pines were nowhere to be found on the Colorado Plateau. Instead, boreal conifers (limber pine, Douglas fir, white fir, and blue spruce) extended 2600–3200 ft downslope of their modern elevational amplitude, beneath which was a sagebrush steppe.

It is interesting to note that ponderosa and pinyon pines may have been invisible to fossil records rather than completely missing from the plateau. The earliest ponderosa pine needles on the Colorado Plateau date between 10,600 and 9800 years ago, but the only available records are from sites above 6500 ft in elevation (Anderson 1993; Betancourt 1990; Weng and Jackson 1999). There is a clear trail of pinyon pine needles in packrat middens that record woodland migration onto the Colorado Plateau at about 11,000 years ago (Allen et al. 1998). It is possible that during the Pleistocene, ponderosa and pinyon pines were surviving unnoticed on the plateau in isolated microhabitat pockets, such as canyon alcoves, and they expanded in range as the climate changed (Anderson 1989; Weng and Jackson 1999).

Table 7. Strengths and limitations of paleoecological proxy data sources.

	Comments	Sensitive to High Frequency (10–100 yr scale) Change	Local Site or Regional Sensor	Continuous Temporal Record	Limitations
Packrat middens	Packrats (*Neotoma* spp.) gather plant and other materials from within about 100 m of their nests. The plants provide food and construction materials that are cemented into their nests with urine, which over time becomes crystallized into a hard shell called amberat.	No	Local	No	• Usually impossible to identify packrat species from fossil middens. • Species diet preferences and collecting behavior vary and species can co-occur in the same area. • Period of time to build midden uncertain. • Middens may be used by other animals. • Most species build middens in rocky areas or cliffs, which are typically more mesic microhabitats with higher plant diversity compared to communities away from cliffs.
Pollen	Pollen grains and spores are extremely resistant to decay and can be identified to plant family, genus, or species.	No	Regional	Yes	The following factors can bias pollen representation of vegetation: • Different dispersal modes (wind or insect) by species. • Physical and chemical processes after deposition. • Differential preservation. • Local to regional vegetation and topography. The major assumption in pollen-based reconstructions is that modern pollen representation is an analog for interpreting fossil pollen spectra. However, past vegetation communities may have no modern correlation.
Charcoal	Fires produce charcoal that is deposited in basins, such as lakes and bogs. Quantifying charcoal particles in sediment cores can be used to examine past fire occurrence.	No	Regional/Local	Yes	• Charcoal production is related to fire intensity and fuels; however charcoal abundance in stratigraphic record cannot be translated to fire severity. • Charcoal transport and deposition in collecting basin is poorly understood; charcoal deposition may continue for several years after fire event, smoothing fire event record. • Methodological isssues.

Table 7 (continued)

	Comments	Sensitive to High Frequency (10–100 yr scale) Change	Local Site or Regional Sensor	Continuous Temporal Record	Limitations
Geology and stratigraphy	Sediment layers and units record periods of erosion and deposition.	No	Regional/Local	No	• Sediments and soils represent a variety of local site conditions and climates. • Sediments and soils can represent relatively long periods of time. • Soils can be difficult to interpret accurately because of the complex nature of their formation.
Tree-ring studies	Tree-ring widths are sensitive to environmental controls.	Yes	Local to Regional with Networks	Yes	• Record fades as number of samples decreases over time. • Loss of large trees with long records from logging and fires. • Imperfect correlation between climate and tree growth—tree-ring widths respond to winter precipitation. • Typically only precipitation reconstructions possible (missing temperature). • Potential distortion from calibration to modern instrumented records, as 20th-century climate may be anomalous relative to earlier times; low spatial resolution of fire area.
Repeat photography	—	Yes	Local	No	• Photos from different seasons could give faulty information on vegetation cover. • Landscape views do not give good idea of amount of bare ground or species composition

Ponderosa pine, pinyon pine, and Utah juniper (*Juniperus utahensis*) first appear in the paleo-record between 11,000 and 10,000 years ago. This shift from boreal forests to pine-dominated forest and woodland likely reflects significant warming and the development of strong summer monsoon activity, about 9000 years ago (COHMAP 1988; Thompson et al. 1993; see Betancourt et al. 1990 for debate on precise timing). The period beginning ca. 9000 years ago signaled the onset of the bimodal storm patterns (winter/spring Pacific storms and southern ocean summer monsoons) that define modern Southwest weather.

The early Holocene warming, considered to be a global event, included soaring temperatures and rapid environmental change. Abrupt climate change is currently a hot research topic (National Research Council 2002; NOAA 2004; Overpeck 1996). Paleoclimate records are revealing that rapid, extreme shifts with temperature increases on the order of 10° C (18° F) or more have occurred several times over the last 100,000 years. During the early Holocene, in the Southwest United States, an increase in forest fires appears to have been contemporaneous with climate and vegetation change (Anderson et al. 2003a; Weng and Jackson 1999).

There is a remarkable consensus from an array of paleoecological records about the major climate inflections over the past 10,000 years, although each site contributes unique histories that differ in the details of timing and duration. At least two significant periods are recognized: an arid middle Holocene period beginning about 9000–7000 years ago and extending to approximately 6000–4000 years ago, and a cool and wet late Holocene starting at about 2000 years ago (Anderson 1993; Blinn et al. 1994; Dean et al. 1996; Weng and Jackson 1999).

New and improved dating techniques and increased resolution of proxy data are filling in the details of (global) Holocene climate connections, revealing climate variability that exceeded any event measured by the instrument-based record of the last 150 years (Overpeck 1996). Many of the climate

events over the last 10,000 years, if repeated today, would have a major impact on regional ecosystems and on their human occupants. Multiple periods of sand dune movement in the Midwest and in northeast Colorado indicate that aridity was accompanied by periods of intense wind (Dean et al. 1996). Middle Holocene sediments are missing from most of the Southwest lake records, which means that the lakes dried out or became very shallow (Anderson 1993; Weng and Jackson 1999). There is one unusually continuous record from Montezuma Well on the Mogollon Rim in Arizona, where diatoms, pollen, and a carbonate profile from a sediment core show that there were several wet and dry intervals during the middle Holocene, with maximum aridity at 7800–6900 years ago and moderately wet conditions 5800–2000 years ago (Blinn et al. 1994).

Climate changes over the last thousand years have been studied using high-resolution data sources, especially tree rings, and within the last few hundred years they are corroborated by historic accounts and instrumented weather records. The last 1000 years are particularly important to the present study, as this period contains no more than three to four generations of our modern forests. Three distinct global climate events are visible: the medieval warm period (900–1300), the little ice age (1400–1900), and the last 110 years. Environmental as well as cultural changes are correlated with these climatic periods; however, as more precise information becomes available, it is becoming clear that the expression and intensity varied greatly from place to place (Hughes and Diaz 1994; Mann et al. 1999).

The two longest tree-ring chronologies in the Southwest are from El Malpais in New Mexico, which extends to 136 BC (Grissino-Mayer 1995), and the bristlecone pine record from the San Francisco Peaks in Flagstaff, Arizona, which starts at AD 581 (Salzer 2000a). The San Francisco Peaks 1419-year bristlecone pine record is a remarkable reconstruction of precipitation and temperature, specifically summer temperatures, a weather parameter that has otherwise been elusive in tree-ring studies. There were 29 extreme wet and 29 extreme dry periods, with an average duration of 10 years (range 5 to 29 years). There were 30 extreme cool and 27 extreme warm periods with an average duration of 16 years (range 5–44 years). The San Francisco Peaks (Salzer and Dean 2004) and El Malpais (Grissino-Mayer and Swetnam 1997, 2000) records correspond on 21 dry periods and 14 wet periods (Table 8), indicating synchronous regional climate intervals.

There are two significant arid periods in the San Francisco Peaks tree-ring record at 1276–1288 and 1292–1300, with an intervening 3-year wet period. The two dry periods combined correspond to the Great Drought (1276–1299), which has been linked to regional scale prehistoric population movements. Another distinct period in the bristlecone pine temperature record consists of three closely spaced cold intervals between 1195 and 1271, which resemble a reconstructed 1200s cold period in southwest Colorado (Petersen 1988). Salzer (2000b) correlated the bristlecone pine cold period to volcanic activity, as recorded by distinct chemical signatures preserved in polar ice caps (Hammer et al. 1980). Multiple or large volcanic eruptions are known to change climate at a decadal scale by increasing the aerosol load in the atmosphere, which reduces the amount of solar radiation received at the earth's surface (Rampino and Self 1984). Based on the San Francisco Peaks tree-ring data, Salzer (2000b) calculated that during one cold period, mean annual temperatures dropped by 1.1–1.6° C (1.98–2.88° F). He has suggested that the cold interval during the twelfth century contributed to the regional abandonment of puebloan sites evident in the Southwest archaeological record.

Analysis of a composite record from 25 climate-sensitive Southwest archaeological tree-ring chronologies and two bristlecone pine tree-ring studies has produced a 1500-year climate record on the Colorado Plateau (Ahlstrom et al. 1995; Dean and Funkhouser 1995). Tree-ring climate reconstructions from wood collected at archaeological sites are usually criticized by climatologists as being biased (e.g. Grissino-Mayer and

Swetnam 2000: 214); however, the inferred wet and dry intervals from the Dean and Funkhouser (1995) composite record generally match other tree-ring based climate reconstructions. The archaeological composite shows that after the AD 900s, the persistent pattern on the Colorado Plateau was a bimodal winter storm/summer monsoon precipitation pattern in the northwest and a unimodal summer-dominant pattern in the southeast. One major period of disruption was identified between 1250 and 1450, a time when climate was more variable and the seasonal patterns of precipitation were unpredictable. The beginning of this period coincides with the critical time of abandonment of large pueblos on the plateau.

Modern weather is generated by atmospheric pressure and temperature differences on land and in the oceans, and the transfer of energy through ocean currents and atmospheric streams. Climate variability is linked to events in the tropical and northern Pacific Ocean that are tracked by the Southern Oscillation Index (SOI) and sea surface temperatures (SST) around the equator (Diaz and Markgraf 2000). The SOI is the standardized difference in sea-level atmospheric pressure between Darwin, Australia, and Tahiti. A negative trend in SOI precedes warming sea surface temperatures in the tropical eastern Pacific Ocean. Warm SST and sustained negative SOI indicate El Niño conditions, whereas cool SST and positive SOI indicate La Niña conditions. This interaction (ENSO, or El Niño–Southern Oscillation) is a high-frequency phenomenon with a period of 4–7 years, although its strength varies. El Niño conditions nudge the Pacific Ocean winter storms south, bringing wet

Table 8. Major climate intervals from El Malpais tree-ring record (adapted from Grissino-Mayer and Swetnam 2000).

Climate Trend	Time Period AD
Wet interval	1000–1400
Below-average precipitation	1400–1790
Above-average precipitation	1790–1992

winters to the Southwest; La Niña conditions result in dry Southwest winters (Cayan et al. 1999).

Hereford et al. (2000) found complicated but significant relationships between Southwest weather records and ENSO. Strong El Niño episodes increased the variability of summer monsoons and the frequency of above-normal winter precipitation, whereas La Niña episodes typically produced normal, relatively low variability warm-season precipitation and typically below-normal winter precipitation (Hereford et al. 2000).

In analyzing the daily records from 97 long-term weather stations on the Colorado Plateau, Hereford et al. (2002) found three distinct periods between 1900 and 2000 (Figure 6): a middle dry period (1942–1977) sandwiched between two multidecadal wet periods (1905–1941 and 1977–1998). The first 5 years of the record (1900–1904) mark the end of an 11-year drought in the Southwest that began in the 1890s (Gatewood 1962). The 35-year dry period between 1942 and 1977 contained the 1950 to 1956 drought, which was the worst drought over the last century in the regional Southwest, and

which caused dramatic vegetation changes in the Southwest (Allen and Breshears 1998). Recent news releases by Dr. R. Webb of the U.S. Geological Survey (USGS) verify that the current drought is more severe than that during the Dust Bowl, and Colorado River flows are currently half those during the 1930s and 1940s (June 10–17, Arizona newspapers).

The integration of paleoecology studies, weather data, tree-ring data, and forest demography studies provides a glimpse of the remarkable climate-synchronized ecological pulse of Southwest forests. Paleo records are invaluable to understanding that Southwest forests have been evolving and adapting to dynamic climate regimes for the past 11,000 years, and that the various ecological and environmental processes are integrated over different temporal and spatial scales. It is clear from this backward view through time that it is not only the cyclical climate patterns that influence forest ecology, but variability, extreme wet and dry intervals, and random events, such as global volcanism, are also key shapers of forest processes.

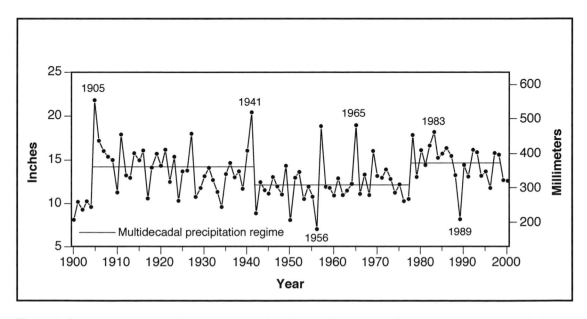

Figure 6. Average annual precipitation recorded at 97 weather stations from 1900 to 2000. Labeled years note extreme wet or dry years, and horizontal lines show average precipitation for three distinct regimes. From Hereford et al. 2002.

The Forest, Climate, Fire, and Insect Connection

Connections to ENSO cycles have been correlated to fire frequency in Southwest ponderosa pine forests, as derived from fire scar tree-ring studies (Grissino-Mayer and Swetnam 1997, 2000; Swetnam and Betancourt 1990). Swetnam and Baisan (2003:173) have emphasized the record of regional fire years; for example, during 1748, large areas burned in almost every mountain range in the Southwest. According to Allen (1998), stand-replacing fires occur throughout the mountain ranges of the Colorado Plateau in the very same season about four times per century. Regional fire years tend to occur during La Niña events and winter/spring droughts, and reduced fire activity corresponds to El Niño wet years (Swetnam and Baisan 2003). Regional fire years also tend to occur during average or dry years that follow one to three wet years; the wet years accommodate the growth of fine fuels (grasses and tree needles) that carry fire between stands of trees (Touchan et al. 1996).

The period 1700 to 1900 is supposedly when the West was "settled" by Anglos, but it is obvious that Native and Hispanic American settlements were already in place. It is nevertheless evident from the case studies presented here that this was a time of intense and extensive resource extraction and land use. This time of transition also encompasses a distinct interval in the 1700s of decreased fire frequency, one that has been attributed to reduced grass cover and the fine fuels necessary to carry fire. This decrease was presumably due to overgrazing by sheep and cattle (Allen et al. 1998; Touchan et al. 1996). The view provided by a network of detailed fire scar chronologies now adds another layer of complexity to the story. In the early 1800s, the Southwest fire regime changed once again, and became characterized by decreased fire frequency and long fire-free intervals, while the fire season shifted from late summer to early spring (Grissino-Mayer and Swetnam 2000).

Swetnam and Baisan (2003:182–184) have suggested that the decrease in fires correlates to a weaker El Niño pattern (dry winter/spring) that may have been amplified by an anomalous cold period between 1740 and 1840. It is interesting that this cold period may have been related to global volcanism (Swetnam and Baisan 2003:183), echoing Salzer's (2000b) theory of a volcanic-caused cold period in the late 1200s. After 1840, livestock grazing intensified in some areas, reaching maximum levels by the late 1880s. Overgrazing provided de facto fire suppression that was reinforced by national fire suppression policies enacted around 1912.

Other forest processes are also likely altered by climatic variability and the ENSO rhythm of wet and dry years. Researchers are beginning to analyze temporal and spatial patterns in forest demography, seedling recruitment, insect outbreaks (Lynch and Swetnam 2003; Swetnam and Lynch 1993), and cycles of masting when cone production peaks. Pulses of ponderosa pine seedlings occurred in many areas between 1800 and 1850 and between 1912 and 1919. The 1800–1850 seedling releases corresponded to a wet cool period (Salzer and Dean 2004) and the longest interval between widespread fires at numerous sites in the Southwest, based on tree-ring studies (Swetnam and Baisan 2003). In a parallel manner, tree mortality from bark beetles increased between 1999 and 2000, from 19,000 acres infected to 750 million acres, as insect populations exploded in response to warm, dry winters and summer droughts (Anhold et al. 2001; U.S. Forest Service 2002). Thus, climate-driven shifts in fire frequencies and pest outbreaks periodically reset the demographic clock for the Colorado Plateau's forests and woodlands, but the recovery time after stand-replacing fires and beetle infestations then depends, to a large extent, on the landscape mosaic established by cultural land uses, including wood cutting, farming, livestock grazing, and hunting and predator control.

Bandelier National Monument
Photo courtesy of Susan J. Smith

The critical need for site-specific management and restoration plans is readily seen on the Colorado Plateau, where the magnitude of both biological and cultural diversity is remarkable for a region this size (Figure 7). The Colorado Plateau is ranked in the top 4 of 109 ecoregions in North America for species richness in several taxonomic groups, and first for unique or endemic biota (Nabhan et al. 2002b; Ricketts et al. 1999a, 1999b). In addition to this great biological diversity, this region contains the greatest number of extant Native American languages of all North American ecoregions, and 29 percent of the lands on the plateau are under tribal control (Nabhan et al. 2002b). What's more, this region offers an exceptionally complete prehistoric and historic ethnobotanical record that reveals cultural interactions with landscape-level biodiversity.

The following case studies demonstrate not only the heterogeneity of the present landscape on the Colorado Plateau, but also the variability in historic land uses and other ecological changes that have shaped the current conditions in each of our study areas. The antiquity and intensity of impacts of human habitation on these landscapes, the timing and intensity of grazing and logging, differences in management practices, and variability in the landscapes themselves have created unique situations in each case. The case studies also provide an opportunity to examine local declines in non-timber forest products or other biodiversity measures. The four study areas were chosen to highlight the range of variability on the Colorado Plateau; each has long Native American occupancy, National Park Service

(NPS) presence, and National Forest Service and/or tribal forestry (e.g., Navajo) management.

Here, we provide information on how to build a site-specific reference envelope for guiding restoration and management practices. Understanding the past shapers of a landscape is an important step in designing a functional management plan.

THE JEMEZ MOUNTAINS AND BANDELIER NATIONAL MONUMENT

The Jemez Mountains, in northern New Mexico, are the southern extension of the Rocky Mountain geographical province. The heart of these mountains is the Sierra de los Valles, with three peaks higher than 10,000 ft, forming a dramatic landscape west of Los Alamos National Laboratory. The Valles Caldera is a prominent landscape feature located just west of the Sierra Crest, and the Rio Grande Valley is 15 miles to the east. Los Alamos and Bandelier National Monument (Figure 8) are situated on the Pajarito Plateau, a broad pediment sloping off the east side of the Jemez Mountains, carved up by canyons into a series of long east-southeast trending mesas. The terrain is volcanic; a stack of dacites, andesites, rhyodacites, and quartz latites formed the mountains. The Pajarito Plateau is cast from a series of massive ash-flow tuffs (ignimbrites; Goff et al. 1996; Griggs 1964; Reneau and McDonald 1996; Smith et al. 1970).

Modern vegetation is a complex mosaic of communities nested at different scales (Allen 2004; Foxx 2002). A broad classification of five main cover types is defined from

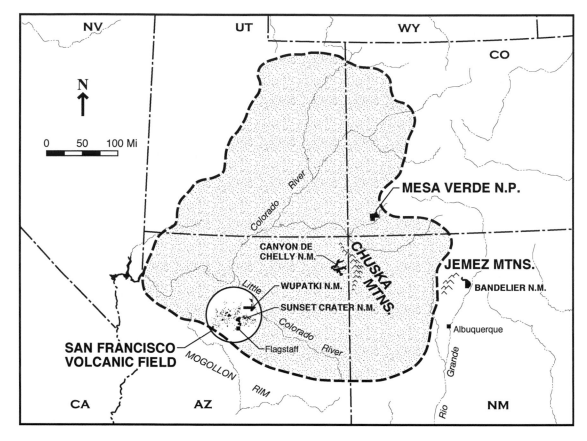

Figure 7. Colorado Plateau showing locations of the four study areas.

the edge of the Pajarito Plateau to the mountain crest: juniper savanna, pinyon-juniper woodland, ponderosa pine forests, mixed conifer forests, and spruce-fir forests (Balice et al. 1997). Boundaries between plant communities are diffuse transition zones moderated by slope, aspect, and the many canyons that carry narrow stringers of higher elevation species through lower elevation mesas. The variety of habitats has created a remarkably diverse flora that has been studied by several researchers (Allen 1989; Foxx and Tierney 1985; Hartman 2003; Jacobs 1989). More than 700 plant species have been documented in Bandelier National Monument alone (Jacobs 1989), which, along with Canyon de Chelly (795 taxa), represents one of the most diverse local floras in the West.

The climate is semi-arid and continental, with cool winters and moderately warm summers. Annual precipitation ranges from 9.8 inches at the lowest elevations to about 35.4 inches at higher elevations (Bowen 1990). Moisture is delivered as snow during the winter months and rain during the fall and spring. The April through late June season is typically dry, and then, between July and September during "monsoons," almost half of the annual precipitation comes during thunderstorms. The overall weather patterns in the Jemez Mountains are highly variable, not only from year to year, but also from season to season. Over a 69-year period, annual precipitation extremes ranged from 7 to 30 inches (Bowen 1990).

Culture History

Of the four landscapes examined here, the Jemez Mountains area has the longest history of continuous human occupation. The 10,000 year long archaeological record is

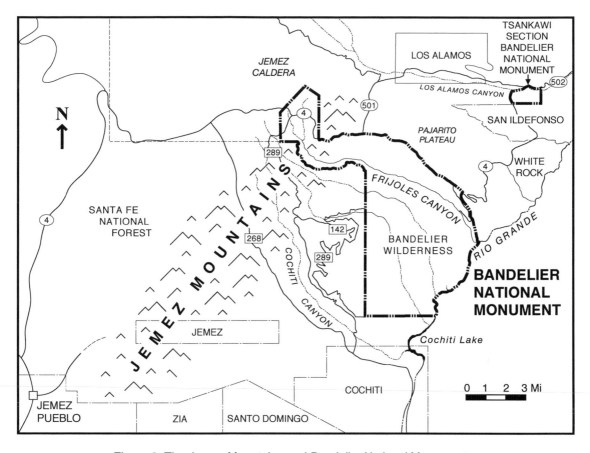

Figure 8. The Jemez Mountains and Bandelier National Monument.

characterized by regional and local shifts in demography and land-use practices, a rhythm of boom and bust, occupation and abandonment. Hunter-gatherer bands occupied sites sporadically in the area beginning around 10,000 years ago. The earliest evidence of farming in this study area is from Jemez Cave, where cob fragments of corn (*Zea mays*) have been radiocarbon dated to more than 3100 years ago (1200 BC; B. Vierra, personal communication 2004). The first extensive farming on the Pajarito Plateau was conducted around AD 1150 by the Pueblo III people. The greatest populations occurred between 1200 and 1500, when a succession of settlement styles coalesced into large agriculture-based pueblos situated in the pinyon-juniper woodlands (Vierra et al. 2002). In the mid-1500s some of the large pueblos were abandoned, possibly due to

climatic change and/or soil depletion, and it is presumed that their occupants moved into communities along the Rio Grande. Today, 14 long-established pueblos are still found in the surrounding watersheds along the major valleys of the Jemez Mountains (Allen 2002a).

Archaeobotanical studies from ruins excavated at Bandelier National Monument and Cochiti Pueblo document a consistent suite of culturally utilized plant resources in the archaeological record (Table 9). In this list are glimpses of plants that are becoming rare in the area today, such as birch (*Betula*) and ash (*Fraxinus*). Ash may have been extirpated from riparian habitats in the Jemez Mountains (Allen 2004), and birch is scarce in New Mexico except for water birch (*Betula occidentalis*), which grows along streams in the Jemez Mountains (Foxx and Tierney

1985; Hartman 2003). There is also one known example of a small, heavily browsed population of bog birch (*Betula glandulosa*) at Alamo Bog in the Jemez Mountains (Allen 2004; Brunner-Jass 1999). Brunner-Jass (1999: 63) suggested that these shrubs are relicts from earlier, more favorable climatic conditions, and that more recent environmental and cultural stresses (e.g., historic overgrazing) may have contributed to the shrinking of wetland and riparian habitats.

More than 200 plant species documented as ethnobotanical resources for Puebloans are present in the Jemez Mountains, based on ethnographic literature and interviews with Pueblo tribal elders (Table 9; Foxx 2002). Using plant summaries in Dunmire and Tierney (1995), Foxx (2002) compared the probable distribution of ethnobotanical resources by plant communities (Table 10); the woodlands and forests of the Jemez Mountains are the main preserves for the range of prehistorically utilized plants.

The Spanish colonized northern New Mexico in the late sixteenth century, and the economic and political power in the region shifted from pueblo farmers to Spanish mission settlements. The Spanish introduced a variety of livestock, including goats, sheep, burros, cows, pigs, and horses (Baydo 1970). They imposed the *encominenda* system of tributary labor on the Indians, and established the *partida* system for settlers. The partida was a share cropper–like contract that involved tending another's herds for a percent of the animals raised. The Pueblo Indians revolted and expelled the Spanish from northern New Mexico in 1680, but the Spanish regained power by 1693, and reorganized the political and economic structure of the region. Land was granted both to Hispanic communities and to individuals who worked the property themselves. Hundreds of small subsistence ranches were developed that remain scattered throughout the Rio Grande Valley and its mountain tributaries (Vierra et al. 2002).

Mexico governed the territory from 1821 to 1846, initiating the era of land grants, during which large tracts of land were transferred to a few wealthy individuals. In 1848, the United States acquired the region as a result of the Mexican-American war, and some but not all of the land grants and corporate holdings in the Jemez Mountains were honored by the U.S. government. In the 1860s, Navajo and Apache Indians were confined to reservations, ending the raiding that had discouraged homesteaders and opening the land to settlers. The railroads reached New Mexico in the 1880s and land use was transformed from subsistence to commercial activities across large areas, with Indian and Hispanic families losing deed to much of their former land (Rothman 1992). The recent history of the Jemez area is marked by the creation of the Los Alamos National Laboratory in 1942, which created almost overnight a new town (White Rock) of primarily white, upper middle class, highly educated professionals building nuclear bombs in the midst of a Hispanic and Pueblo Indian world (Allen 2002b).

Land-Use History

Farming has been a tradition in the Jemez Mountains and northern New Mexico since the AD 1100s. By the 1300s, Anasazi farmers were using sophisticated dry-land farming techniques in 1–10 acre plots, managing and conserving all manner of surface water (Toll 1995). By the 1700s, Spanish subsistence farmers and herders were setting up small settlements, and consolidating former fields into larger pastures and orchards. However, Indian raiding from the 1700s through the 1800s discouraged permanent Spanish settlements in the mountains, where land use became primarily seasonal and relatively light in intensity (Allen 2004). The 1790 Santa Fe census listed the population as more than 2500 souls, and almost everyone listed their profession as farmer (Dorman 1996). The Spanish imported their Old World, or Mediterranean cuisine, flora, and acequia system to New Mexico and the local diet changed to include wheat, lentils, chick peas, watermelons, cantaloupes, cucumbers, cabbage, lettuce, carrots, artichokes, garlic, onions, plums, apricots, and peaches (Lopinot 1986). From the 1880s to the 1930s, the region contained Anglo, Indian, and

Table 9. Plant species recovered from archaeological sites in the eastern Jemez Mountains.

Scientific Name		Common Name	Probable Uses
Agavaceae	*Yucca* sp.[2]	Yucca	Food, fiber, soap
Amaranthaceae	*Amaranthus* sp.	Pigweed	Food
Anacardiaceae	*Rhus trilobata*	Skunkbush	Food
Asteraceae	*Artemisia tridentata, A.* sp.	Sagebrush	Fuel, food
	Ericamera nauseosa, E. sp.	Rabbitbrush	Fuel, dye
	Helianthus sp.[1,2]	Sunflower	Food
Betulaceae	*Betula* sp.	Birch	Fuel, construction, baskets
Cactaceae	*Coryphantha* sp.	Pincushion cactus	Food
	Fouquieria splendens, F. sp.	Ocotillo	Construction, tools
	Echinocereus sp.	Hedgehog cactus	Food
	Opuntia imbricata	Walking cane cholla	Food, tools, living fence
	Opuntia sp.[2]	Cholla & prickly pear	Food, medicine
Capparaceae	*Cleome* sp.[1]	Beeweed	Food
Chenopodiaceae	*Atriplex* sp.	Saltbush	Fuel
	Chenopodium sp.	Goosefoot	Food
Cucurbitaceae	*Cucurbita pepo, C.* sp.[1]	Squash	Food
Cupressaceae	*Juniperus* sp.	Juniper	Fuel, construction, food
Cyperaceae	*Carex* sp.	Sedge	Food, baskets
Euphorbiaceae	*Euphorbia* sp.	Spurge	Food, medicine
Fabaceae	*Phaseolus* sp.[1]	Bean	Food
Fagaceae	*Quercus* sp.	Oak	Fuel, construction, food
Malvaceae	*Gossypium* sp.[1]	Cotton	Fiber, food
	Shaeralcea sp.	Globemallow	Food, medicine
Oleaceae	*Fraxinus* sp.	Ash	Fuel, construction
Pinaceae	*Pinus edulis*	Pinyon pine	Fuel, construction, food
	Pinus ponderosa	Ponderosa pine	Fuel, construction, food
Poaceae	*Phragmites australis*	Common reed	Fuel, construct'n, smoking
	Sporobolus sp.	Dropseed	Fuel, textile
	Zea mays[1]	Corn, maize	Food, fuel, ceremony
Polygonaceae	*Eriogonum* sp.[2]	Buckwheat	Food, medicine, ceremony
Portulacaceae	*Portulaca* sp.	Purslane	Food
Rosaceae	*Cercocarpus montanus, C.* sp.	Mountain mahogany	Fuel, tools, dye
	Prunus virginiana	Common chokecherry	Fuel, food
Salicaceae	*Populus* sp.	Poplar, aspen	Fuel, construction, food
	Salix sp.	Willow	Fuel, construction, baskets
Solanaceae	*Datura* sp.	Jimson weed	Medicine, ceremony
	Lycium sp.	Wolfberry	Fuel, food
	Nicotiana attenuata	Tobacco	Medicine, ceremony
	Physalis sp.	Groundcherry	Food
Typhaceae	*Typha* sp.	Cattail	Food, fiber, ceremony
Ulmaceae	*Celtis reticulata*	Hackberry	Fuel

[1]Cultivated species.
[2]Taxa identified only from pollen.
Sources: Allen 2004; Brunner-Jass 1999; Foxx 2002; Smith 1998; Vierra et al. 2002.

Table 10. Plant uses and numbers of plant species used from Jemez Mountain plant communities (Foxx 2002).

Uses	Riparian	JS	PJ	Pipo	MC
Medicinal (n = 148)	18	82	111	73	35
Food (n = 108)	23	41	77	56	30
Implements (n = 28)	4	14	20	15	6
Coloring, tanning (n = 37)	6	19	24	16	6
Construction (n = 16)	6	7	9	8	4
Smoking (n = 13)	0	8	11	3	9
Cordage (n = 6)	2	3	2	2	1

JS = juniper savannah, PJ = pinyon-juniper, PiPo = ponderosa pine, MC = mixed conifer.

Hispanic subsistence and commercial fields, orchards, and pastures. In 1937, about 6 square miles on the Pajarito Plateau was farmed by 35 Anglo and Hispanic homesteads (Foxx et al. 1997). Figure 9 shows an aerial view of dry-farmed bean fields on the present-day Los Alamos town site.

Though perhaps inconsequential on a landscape level, livestock first set hoof in New Mexico well before the Spanish settlements of the late sixteenth century. Mexican Indians in the 1500s were trading for horses with the Spanish and exporting livestock north (Baydo 1970). Large numbers of livestock, especially sheep, were imported during Spanish colonization in the early 1600s. Sheep were the keystone of a Spanish subsistence economy, and became so for many Pueblos as well (Carlson 1969). Cattle only surpassed sheep in numbers in the 1800s when Anglos started settling in New Mexico. Economic policies and laws passed by the Texas legislature in 1879 and 1883 restricted cattle grazing on depleted Texas ranges, and cattlemen moved over to New Mexico (Foxx 2002:23). The most abusive overstocking started in the 1880s when the railroads opened the "frontier," and extended into the early 1900s. By 1895 virtually every acre of accessible grassland in New Mexico was overstocked with sheep or cattle at four times or more of what the range could carry (Foxx 2002: Appendix 8). Grazing was not excluded in Bandelier until the Park Service took over in 1932. Even after domestic animals were removed, feral burros introduced by the Spanish still overgrazed prime habitats and impacted streams and springs until they were removed in the 1970s (Allen 2004).

Foxx (2002:23) has provided an interesting history of one of the Pajarito Plateau grants; it is an example of what happened throughout the New Mexico Territory during the era of land grants:

In 1742 Pedro Sanchez was given a grant, that later became known as the Ramon Vigil Grant. For over 100 years, the land remained within the Sanchez family and was primarily used for seasonal grazing. In 1851 Antonio Sanchez sold 8 of the 10 land shares to Ramon vigil. Twenty-eight years later, Ramon Vigil sold the land to Thomas Aquinas Hayes, an Irish priest, for $4000. Within a month Hayes had sold it for $16,000, later buying it back for the same price. On June 13, 1884, Padre Hayes sold the grant to Winfield Smith and Edward P. Shelton of Cleveland Ohio for $100,000. These sales took the land permanently out of Spanish hands, ending an epoch of isolation, years of subsistence farming, and pastoral grazing. From approximately 1885 through 1887, the Ramon Vigil Grant was rented to a Texas cattleman, W. C. Bishop, who ran 3000 head of cattle on 32,000 acres. This was severe overstocking by a factor of about 10—a liberal range estimate for the grant's carrying capacity is 100 acres/cow/year. In 1897, the timber rights on the grant were sold to H.S. Buckman, who logged the area until 1903. He built the town of Buckman on the Rio Grande as a railway station and constructed a road linking the rail stop to his sawmill. A newspaper article of December 1903 speculated that Buckman cut 36 million board feet on the 32,000 acre grant. Throughout the 1900s and 1940s areas adjacent to the Ramon Vigil Grant yielded 17,319,000 board feet of saw log timber.

Figure 9. Aerial photograph of fields in the Los Alamos area. Photo courtesy of National Archives and Records Administration.

Fire History

Before the 1880s, there were frequent widespread lightning fires in the Jemez Mountains, but few "Indian" fires were recorded (Allen 2002a). The pre-1900 fire return intervals at most fire scar study sites ranged from less than 10 to 20 years, with more frequent fires in ponderosa pine forests (Allen 2004; Grissino-Mayer and Swetnam 1997). In the wetter, higher elevation mixed conifer forests, fires tended to be crown fires that killed small patches of trees (Allen 2004). Fire patterns in the pinyon and juniper woodland are more complicated. Baker and Shinneman (2004) and Floyd et al. (2004) claim that there are no reliable data to support the idea that low-severity surface fires were common in pinyon-juniper woodlands, and they suggest that these woodlands are characterized by long-rotation, on the order of 400 years or more, high-severity fires.

Tree-ring studies in the ponderosa pine forests in the Jemez Mountains show that from 1600 to the 1700s, fire regimes were characterized by high-frequency, low-intensity surface fires, with a sharp decrease in fires beginning about 1750 (Touchan et al. 1996). This presumed fire gap may correlate with the first overgrazing by sheep in the region, which reduced grass and other fine fuels (Touchan et al. 1996). An alternative theory, proposed by Swetnam and Baisan (2003:182–184), is that global El Niño climate trends and possibly major volcanic eruptions in the early 1800s caused a distinct cooler period, which dampened fire activity.

Whatever the cause, suppressed fire regimes from 1750 to 1790 were succeeded by de facto fire suppression from livestock overgrazing and logging from the 1880s to 1900s. The net effect of human and livestock impacts and the early 1900s policy of fire suppression was an unnatural, reduced fire frequency (Allen 2002b; Allen et al. 1998). In the Southwest, including in the Jemez Mountains region, there were at least two significant wet intervals in the 1900s, one from 1905 to 1941 and a second from 1978 to 1998 (Hereford et al. 2000). Forests responded with surges in growth that built up fuels to dangerous levels. There have been catastrophic crown fires in the past 20 years; more than 80,000 acres have burned on the east slope of the Jemez Mountains since the 1977 La Mesa fire. By far the largest and most dramatic was the 2000 Cerro Grande fire on the Pajarito Plateau, which burned 44,000 acres and 354 homes, costing more than 5.1 million dollars (Griggs 2001).

Environmental Consequences of Cumulative History

The complex and at times intensive land uses in the Jemez Mountains initiated a cascade of environmental responses. Local vegetation communities have undergone major changes over the past two centuries from the cumulative and intertwined effects of climate shifts, livestock grazing, farming, logging, development, and fire suppression (Allen 1989, 2002b, 2004; Allen et al. 1998). The tempo of change is apparently increasing with the pressures of modern development. Allen (2002b) documented that human-altered landscapes in 1935 accounted for 0.9 percent of 212,000 acres, compared to 6.3 percent in 1981, which includes Los Alamos National Laboratory, the towns of Los Alamos and White Rock, the Cochiti Lake reservoir, assorted golf courses, pumice mines, and a ski area.

Tree densities in mixed conifer and ponderosa forests have increased approximately tenfold over what they were during pre-1900 conditions (Swetnam et al. 1999), escalating fire hazards to extreme conditions while reducing the cover and diversity of understory shrubs and herbaceous plants. At Bandelier National Monument, Jacobs et al. (2002) consider the pinyon-juniper woodlands to be a fire-intolerant community that in the past may have been limited to fire-proof sites on rocky, shallow soils. More than 150 years of intensive land uses and fire suppression have favored expansion of these woodlands and closing of crown canopies in older stands (Jacobs et al. 2002; see Baker and Shinneman 2004 and Floyd et al. 2004 for alternative interpretations). At Bandelier, pinyon and juniper recruitment increased exponentially after 1880, perhaps as a consequence of local fire suppression (Julius 1999).

As canopies closed, herbaceous ground cover was lost and the amount of bare ground increased. In the Jemez Mountains, exposed tuff and volcanic substrates and shallow soils have become prone to high erosion rates. Allen has documented 58–84 percent bare ground under pinyon-juniper canopies at Bandelier National Monument (Allen 2004). In the Frijolito watershed at Bandelier, sediment is eroding at approximately 2 mm per year, which at this rate would strip all of the intercanopy soil from the watershed within 100–200 years (Allen 2004). In one thunderstorm event at Bandelier, Allen (2002b:242) documented more than 1040 artifacts (mostly potsherds) being moved into a sediment trap that drained only about 11,000 sq ft of gentle hillslope. Loss of the soil means reduced soil organic matter, plant nutrients, water storage capac-

ity, and net primary productivity. As Allen (2002b: 243) put it, "the park's biological productivity and cultural resources are literally washing away."

Expansion of trees into former grasslands and meadows, and the closing of tree canopies, has had a hand in homogenizing the landscape and decreasing biodiversioty in the Jemez Mountains. Between 1935 and 1981, montane grassland communities were reduced by 55 percent across 247,000 acres of the Jemez Mountains due to the expansion of conifer populations (Allen 1989, 1998). In the Valles Caldera, blue spruce is expanding into wet meadows (Allen 2002b: 243). The closure of the pinyon and juniper woodland canopy is thought to have resulted in the loss of a significant intercanopy grass and forb component (Jacobs et al. 2002).

Exotic species are becoming established over large areas and are taking over significant niches, especially the increasingly limited riparian and wetland communities (Allen 2004). At Bandelier National Monument, approximately 17 percent of the park flora consists of introduced plants (Brian Jacobs, botanist, Bandelier National Monument, unpublished data), and at least 13 exotic species were set loose in the Jemez region *before* 1800, well in advance of fire suppression and overgrazing (Foxx 2002).

At Bandelier National Monument in the pinyon and juniper woodland, a large-scale paired watershed study was initiated in 1996 to document a restoration prescription (Jacobs et al. 2002). One 100-acre area was monitored as a control and an adjacent 100-acre area was treated by thinning two thirds of the trees and scattering the slash onto the exposed soil of the newly opened intercanopy areas. The slash dispersal was a key part of the treatment to enhance microsite conditions for plant seedlings to establish; the slash moderated temperature extremes by providing shade and intercepting runoff, increasing infiltration, which both enhanced soil moisture available to plants and decreased erosion. Results after 3 years are dramatic (Jacobs et al. 2002). Grasses have increased from 17 percent of the relative canopy cover to 27 percent and subshrubs from 2 to 5 percent. Grass species composition changed over the course of the 3 years from blue grama grass (*Bouteloua gracilis*) dominant to blue grama and little bluestem (*Andropogon* sp.) co-dominant, along with increases in mutton grass (*Poa fendleriana*; Jacobs et al. 2002:42).

An exclosure within the treated plot was also monitored to test responses when ungulate grazers (both wild and domestic) are restricted. The two most notable differences were that in the exclosure, wavy leaf oak was dominant, but outside the exclosure broom snakeweed was dominant. Grass canopy cover within the exclosure doubled from 5 to 10 percent cover in the first 2 years after treatment, whereas grass cover outside the exclosure did not exceed 10 percent cover until the end of the third year. Subjective comparisons between the exclosure and the watershed suggested that outside the exclosure, cool-season grasses (mutton grass and June grass, *Koehleria macrantha*) were being preferentially grazed to the root crown in spring and fall, with some mortality (Jacobs et al. 2002:49).

The north end of Mesa Verde National Park
Photo courtesy of Northern Arizona University

Mesa Verde National Park (Figure 10) is in the southwestern corner of Colorado, in Montezuma County. The 52,073-acre park was established June 29, 1906 to preserve the many cliff dwellings on the mesa. The park lies on the Mesa Verde cuesta (Figure 11), an erosional remnant tableland elevated above the surrounding valley floor by about 2000 ft, with elevation highest in the north (8500 ft) at Park Point, and decreasing gradually to the south. The mesa is a labyrinth of narrow lateral canyons running north-south, and is drained by the Mancos River and its tributaries. The park is bordered by the Mancos River to the east, the Ute Mountain Ute Indian Reservation to the south and west, and Montezuma Valley to the north (Torres-Reyes 1970).

The bedrock geology of the mesa is dominated by sedimentary formations: the Mancos and Menafee Shales and the Cliff House and Point Lookout Sandstones (Griffitts 2003). The soils in the park range from productive, red eolian soils in the higher elevations to thick deposits of colluvium in the canyon bottoms and fans (Ramsey 2003). Variability in soils supports isolated pockets of rare plant communities across the mesa (Floyd and Colyer 2003).

Mesa Verde receives 18.2 inches of precipitation annually, which is significantly higher than nearby areas such as Hovenweep National Monument, which receives 11.5 inches and the city of Cortez, Colorado, which receives 13.3 inches (Erdman 1970:3; Ramsey 2003:214). The higher precipitation is likely due to the higher elevation of the cuesta, above the surrounding areas. Most of this precipitation occurs as snowfall in winter months; summer precipitation usually

occurs between July and September as short, intense thunderstorms (Omi and Emrick 1980).

Mesa Verde, a cold, middle latitude, semi-arid steppe (Trewartha 1954), supports the largest natural reserve of the Upper Sonoran/Sierra Madrean Complex left in the world (Thomas et al. 2003). Vegetation is co-dominated by a pinyon-juniper woodland and a scrub community, or petran chaparral, dominated by Gambel oak, mountain mahogany (*Cercocarpus montanus*), sagebrush (*Artemisia tridentata*), and rabbit brush (*Ericamera nauseosus*; Floyd et al. 2003d). There are montane meadows of native grasses and an estimated 495 acres of wetland and riparian habitat in the park, supporting communities dominated by willow (*Salix* spp.) and cottonwood (*Populus fremontii*; Floyd-Hannah and Romme 1995; Thomas et al. 2003). Mesa Verde's pinyon-juniper woodlands are some of the oldest and largest on the Colorado Plateau, with trees that are 400 years old or more in many stands (Erdman 1970; Floyd 2003; Floyd et al. 2003a). There is at least one ancient Utah juniper that is 1350 years old (Floyd et al. 2003a). Such older pinyon-juniper habitats include much downed and standing dead wood, which creates shelters and other opportunities for understory plant and animal life (Floyd et al. 2003d). They are also home to some plants that may be endemic to this habitat type but do not occur in younger stands, even at other locations on Mesa Verde (Floyd and Colyer 2003).

Overall species diversity is very high in the park. This diversity may be attributable to its protected status and large wilderness area, as well as to the relatively intact wild-

46

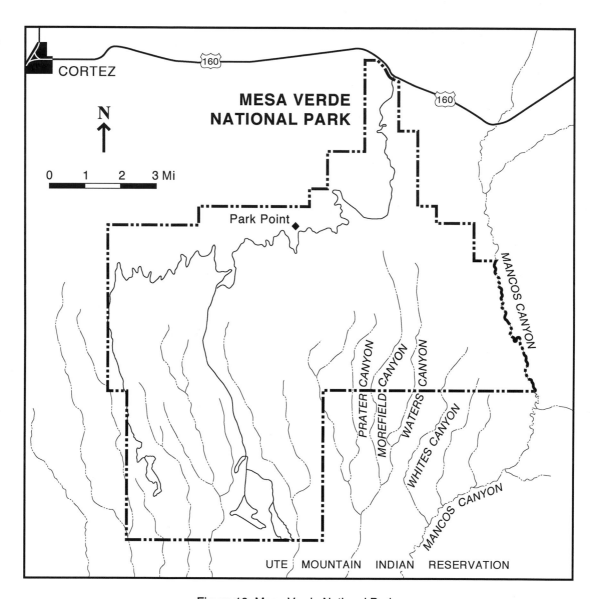

Figure 10. Mesa Verde National Park.

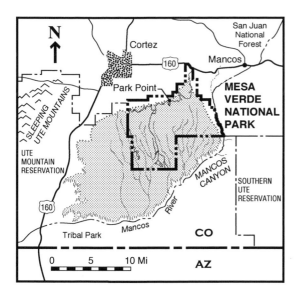

Figure 11. Mesa Verde cuesta, Colorado.

life corridors running into the park; it may also be due to its rich and varied environmental and cultural history. There are 627 native vascular plant species and 87 exotics identified in the park (Thomas et al. 2003).

Culture History

The first human occupation of the Four Corners region around Mesa Verde was by Paleoindian hunters at least 8500 years ago, though there are no known sites of this period on Mesa Verde (Lipe and Pitblado 1999). This nomadic hunting tradition spanned a time of climatic change at the end of the Pleistocene, when conditions generally began to be warmer and drier. Around 8000 years ago, at the beginning of the Archaic period, human groups became more reliant on local wild animal and plant resources. The pinyon-juniper woodlands of the Colorado Plateau were probably established at about this time as well (Adams and Petersen 1999). No later than 7500 years ago, as the Early Archaic period began, a substantial change in subsistence strategies occurred; remains from the Early Archaic show a stronger reliance on wild plant foods. During a subsequent mid-Holocene warm period with increased precipitation

around 6000 BP (though this was probably not generalized across a large area), human populations apparently increased; maize agriculture probably became nutritionally important in this region no later than 400 BC, though its initial appearance as determined archaeologically is somewhat earlier (Lipe and Pitblado 1999).

The culture history of Mesa Verde from AD 700 to 1300 is dominated by Puebloan farmers (Lipe and Varien 1999a, 1999b; Wilshusen 1999). These farmers most often lived in cliff house villages and pueblos near their agricultural fields, which were generally located in the colluvial fans and valley bottoms, though a variety of habitation sites have been identified. Changes in settlement patterns over time have been correlated with climate change as recorded by tree-ring chronologies (Adams and Petersen 1999; Berry 1982). The final Puebloan (Pueblo III, AD 1150–1300) occupation of Mesa Verde was characterized by an intensification of agriculture (including water control systems), decreased reliance on wild resources, and accelerated population aggregation (Wilshusen and Towner 1999). These factors, in addition to a generally cooler and wetter climate followed by a prolonged drought (Van West and Dean 2000), apparently led to the abandonment of this lifeway at Mesa Verde around 1300. In the years prior to the abandonment, evidence indicates an overall decrease in human health (particularly of women and children) and increase in warfare (Wilshusen and Towner 1999). There is evidence of deforestation in an analysis of fuelwood used during this period on nearby Grass Mesa (Kohler and Matthews 1988), though whether or not Mesa Verde experienced similar reductions is controversial, and it could have been caused by climatic rather than cultural changes (M. Floyd, personal communication 2004). In a rainfall/yield-based simulation of agricultural production and storage potentials from 652 to 1968, the drought-ridden period between 1276 and 1299 was shown to have "the worst shortfalls to have plagued southwestern Colorado's Anasazi" (Burns 1983). Mesa

Verde was virtually abandoned around 1300 and remained unoccupied until about 1500, with no evidence of even seasonal use (van West and Dean 2000; Wilshusen and Towner 1999).

The first evidence for reoccupaton of Mesa Verde was left by Utes (Numic speakers) and proto-Navajo/Apache bands (Athapaskan speakers) around 1500 (Wilshusen and Towner 1999). Their early archaeological sites show a reliance on wild resources, supported by the remains of deer, elk, bison, and other small and medium sized game. Occupants also collected goosefoot (*Chenopodium* spp.), grasses, cacti, pinyon nuts, serviceberry (*Amelanchier utahensis*), and other fruits, nuts, and seeds, which have been recovered from the few archaeological habitation sites known for this time period on the mesa (Brown 1996; Wilshusen and Towner 1999).

By the seventeenth century, Navajo subsistence was a mix of hunting, upland gathering, and seasonal farming, with significant storage of corn (Wilshusen and Towner 1999). After the Pueblo Revolt in 1680, the Utes acquired horses from the Spanish and began applying pressure to the Navajos, driving them across the San Juan River to the south.

Although the first known European incursion into Mesa Verde occurred in 1765 with the Don Juan Maria de Rivera expedition, there was very little exploration or contact until the end of the Mexican War of 1848. Even until the late 1800s, the rugged landscape kept many Europeans from coming to the area, which remained primarily Ute territory until about 1881.

Administrative History

There was very little governmental involvement at Mesa Verde until establishment of the park in 1906 to showcase and protect its many spectacular cliff dwellings. Prior to that time it had been exclusively Ute territory; it was simply acknowledged as Ute territory by the Spanish and Mexican governments until the Mexican War. After the war, the U.S. government made its first

treaty with the Utes in 1849, and the first reservation in southern Colorado was established in 1880 (www.utemountainute.com/chronology.htm). This reservation area was reduced drastically upon establishment of the park in 1906, causing so much tension with the Utes that the first park superintendent brought in Navajos to work there. In 1913 there was an adjustment to the boundary of the park to include a significant portion of the area containing ruins (which was located on Ute land), in exchange for the Ute Mountain portion of the park (Torres-Reyes 1970), but the Utes still maintain that the land was taken from them illegally (Burnham 2000). Boundary disputes continue to this day, as do Ute–NPS conflicts over hunting, livestock, fire management, and other resource issues (Burnham 2000).

In recognition of its spectacular archaeological resources, much of Mesa Verde National Park was designated a wilderness on 20 October 1976. It was designated a World Heritage site on 6 September 1978.

Land-Use History

The first herd of cattle was driven onto the mesa in the early 1870s when cattle were brought west, and Ute families were paid to tend them during the spring and summer months on the mesa; also around this time, herds of cattle and sheep were driven through the area on their way from New Mexico to Mormon colonists in Utah (O'Rourke 1980). Beginning in the late 1870s and lasting until around 1935, several homesteads were established and large herds of cattle, sheep, and horses were grazed within what are now the park boundaries. In addition 16 wells or pipe springs were dug by homesteaders, which had important impacts on the hydrology of the canyon bottoms. These wells have since been capped (M. Colyer, personal communication, 6 Oct 2003). Even after the establishment of the park, park superintendents and concessionaires inappropriately grazed large herds of their own cattle in the park (M. Colyer, personal communication 2003; Smith 2003). In his administrative history of the park,

Torres-Reyes indicated that "by the time the park was established Mesa Verde was already over grazed ... A minimum of 2,150 cattle and ... 2,400 sheep were grazed seasonally on park lands without permits or control from 1906 through August 1911" (Torres-Reyes 1970:211–214). Although grazing on park lands (not including inholdings within park boundaries) was made illegal in 1908, superintendents soon realized that, in the absence of fencing, it was impossible to prevent cattle from entering park lands. In 1911 grazing leases were issued in the park to four permittees for a total of 1165 cattle (National Park Service 1942), but limits were typically not enforced, and "as a rule [lease holders] ran all the cattle and sheep they pleased, regardless of the number stipulated in the permits" (Torres-Reyes 1970:213). Although livestock roamed across the mesa, grazing was heavily concentrated—and most destructive—in the canyon bottoms, particularly in Morefield, Whites, and Prater Canyons where homesteads were established, and where private inholdings persisted after the establishment of the park.

When archaeologist Jesse Nusbaum became superintendent in 1921, he was greatly disturbed by the damage to archaeological sites caused by livestock (which seemed to be attracted to the more fertile soils around the sites) as well as erosion and damage to roads and trails. He began a policy of phasing out grazing, and legal grazing on park land (not private inholdings) ended for the most part in 1927, though stray cattle and feral horses continued to graze in the park until around 1960 (M. Colyer, personal communication 2003; Smith 2003). Private inholdings were all either donated to the park or purchased by 1945, with the exception of some later purchases along the Mancos River, the last of which was finalized in 1990 (M. Colyer, personal communication 2003).

After the exclusion of cattle and sheep, there was "good recovery" of springs, streams, and vegetation. The range and canyon bottom lands recovered so well that they became "a veritable showplace for scientific and educational institutions, and personnel throughout the Southwest interested in range recovery and control" (Superintendent's Annual Report 1946). The park was the site of a range training school sponsored by the U.S. Forest Service in 1946 that focused on changes evident in two of the last inholdings to come under park ownership, Morefield and Prater Canyons (Woodhead 1946). In areas documented by repeat photography in 1935, 1942, and 1946 there is evidence of a change from very near 100 percent bare ground to nearly 100 percent cover (mostly clover in many places) within 8 years, followed by mostly grasses and native herbs within 11 years (Figure 12); Woodhead 1946). Recovery of water resources was also dramatic. Measurements of the water table, taken in wells in those canyons in 1935, 1942, and 1946, revealed that the water table in Morefield Canyon had risen 17 ft and the water table in Prater Canyon had risen 9 ft between 1935 and 1946, in spite of a severe drought (Woodhead 1946). U.S. Forest Service officials said that Mesa Verde provided the "finest example of range recovery, and of normal range conditions in the pinyon-juniper type to be found in the Southwest" (Superintendent's Annual Report 1946). At that time, grazing had been excluded from the studied areas for 11 years. Now that grazing has been excluded for more than 65 years, Mesa Verde National Park is an even more important area for studying the effects of release from domestic livestock grazing.

The gathering of wild resources, though prohibited in the park now (with two exceptions), was vitally important to the native inhabitants of the mesa, and was later important to Navajo, Hispanic, and Anglo residents of the region. Archaeobotanical remains at prehistoric habitation sites attest to the abundance and cultural use of wild food, fiber, fuel, and construction resources on the mesa (Table 11). Many native wild food resources are plentiful, including pinyon nuts, cactus pads, cactus fruits and other fruits, acorns, seeds of many grasses and forbs, and a variety of herbs and roots,

Figure 12. Repeat photography in Prater Canyon, near Prater homestead and windmill. A. 1935; B. 1942; C. 1944. Photos courtesy of Mesa Verde National Park.

Table 11. Plant species recovered from archaeological sites in the Mesa Verde region.

Scientific Name		Common Name	Probable Uses
Aceraceae	*Acer* sp.	Maple	Wood, fuel
Agavaceae	*Yucca* sp., *Yucca baccata*	Yucca, banana yucca	Food, fiber
Amaranthaceae	*Amaranthus* sp.	Pigweed	Food
Anacardiaceae	*Rhus aromatica*	Fragrant sumac	Food, fuel
	Rhus trilobata	Skunkbush	Food, fuel
Apocynaceae	*Apocynum cannabinum*	Indian hemp	Fiber
Asclepiadaceae	*Asclepias* sp.	Milkweed	Fiber
Asteraceae	*Artemisia tridentate, A.* sp.	Sagebrush	Fuel, wood, fiber, food
	Ericamera sp.	Rabbitbrush	Fuel, wood, fiber
	Helianthus sp.[1]	Sunflower	Food
	Iva sp.[1]	Sumpweed, marsh elder	Food
Boraginaceae	*Lithospermum ruderale*	Grommel, stoneseed	Food, medicine
Cactaceae	*Opuntia phaeacantha, O.* sp.	Tulip prickly pear, prickly pear	Food
Capparaceae	*Cleome serrulata*,[1] *C.* sp.	Beeweed	Food
Chenopodiaceae	*Atriplex* sp.	Saltbush, shadscale	Food, fuel
	Chenopodium sp.	Goosefoot	Food
	Sarcobatus vermiculatus	Greasewood	Wood, fuel
Cornaceae	*Cornus sericea, C.* sp.	Dogwood	Wood, fuel
Cucurbitaceae	*Cucurbita pepo, C.* sp.[1]	Squash	Food
	Lagenaria siceraria[1]	Gourd	Container
Cupressaceae	*Juniperus* spp.	Juniper	Food, wood, fuel, fiber
	J. osteosperma	Utah juniper	Food, wood, fuel, fiber
	J. scopulorum	Rocky Mountain juniper	Food, wood, fuel, fiber
Cyperaceae	*Scirpus* spp.	Bulrush	Food, fiber
Elaeagnaceae	*Shepherdia argentea, S.* sp.	Buffaloberry	Food, fuel
Ephedraceae	*Ephedra* sp.	Mormon tea	Medicine
Fabaceae	*Phaseolus vulgaris, P.* sp.[1]	Bean	Food
Fagaceae	*Quercus gambelii*	Gambel oak	Food, wood, fuel
Hydrangeaceae	*Fendlera rupicola, F.* sp.	Fendler bush	Wood, fuel
Lamiaceae	*Scutellaria* sp.	Skullcap	—
Liliaceae	*Allium* sp.	Wild onion	Food
Loasaceae	*Mentzelia albicaulis*	Whitestem blazingstar	—
Malvaceae	*Spaeralcea* sp.	Globemallow	Food
Pinaceae	*Abies concolor*	White fir	Wood, fuel, pitch
	Pinus spp.	Pine	Wood, fuel, food, pitch
	Pinus edulis	Pinyon pine	Wood, fuel, food, pitch
	Pinus ponderosa	Ponderosa pine	Wood, fuel, food, pitch

Table 11 (continued)

Scientific Name		Common Name	Probable Uses
Pinaceae (continued)			
	Pseudotsuga menziesii	Douglas fir	Wood, fuel, pitch
Poaceae	*Achnatherum hymenoides*	Indian ricegrass	Food
	Phragmites australis	Common reed	Fiber, fuel
	Poa fendleriana	Muttongrass	Food
	Zea mays[1]	Maize, corn	Food, fuel
Polygonaceae	*Eriogonum* sp.	Buckwheat	Food
	Polygonum douglasii, P. sp.	Knotweed	Food
	Rumex sp.	Dock	Food
Portulacaceae	*Portulaca oleracea, P.* sp.	Purslane	Food
Rosaceae	*Rosa* sp.	Rose	Wood, fuel, food
	Amelanchier utahensis, A. sp.	Serviceberry	Wood, fuel, food
	Cercocarpus montanus, C. sp.	Mountain mahogany	Wood, fuel, dye
	Peraphyllum ramosissimum	Squaw apple	Food, wood, fuel
	Prunus areniaca[1]	Apricot	Food
	Prunus virginiana, P. sp.	Chokecherry	Food, wood
	Purshia tridentata, P. sp.	Cliffrose	Wood, fuel
	Rubus sp.	Blackberry	Food
Salicaceae	*Populus* spp.	Cottonwood	Wood, fiber
	Salix spp.	Willow	Wood, fiber
Scrophulariaceae	*Cordylanthus* sp.	Bird's-beak	—
Solanaceae	*Datura wrightii, D.* sp.	Jimson weed	Medicine, ceremony
	Lycium pallidum	Wolfberry	Food, wood, fuel
	Nicotiana attenuata	Tobacco	Medicine, ceremony
	Physalis hederifolia, P. sp.	Groundcherry	Food
	Solanum jamesii	Wild potato	Food
Typhaceae	*Typha* sp.	Cattail	Food, fiber
Ulmaceae	*Celtis* sp.	Hackberry	Wood, fuel, food
Verbenaceae	*Verbena* sp.	Vervain	—

[1]Cultivated species.
Sources: Colyer 2003; Cordell 1994; Kohler and Matthews 1988; Matthews 1987; Stiger 1979.

as well as animal foods such as deer, fish, small mammals, and amphibians. Fibers needed for clothing, sandals, and basketry were also abundant, and especially useful were the straight, flexible young shoots from shrubs (often plentiful after fire in the petran chaparral zone), and yucca fibers. Fuel and construction materials on the mesa would have been plentiful and easy to gather after fire in the higher elevations. In fact, fire would have increased the available resources dramatically (Adams 2002), and indeed, after recent fires, stone-rimmed terraces built in prehistoric eras have once again become covered with wild onions (*Allium macropetalum*), sego lilies (*Calochortus nuttallii*), goosefoot, and other edible herbs. The Utes relied heavily on hunting and the gathering of wild plant resources for food. The Utes also used the cambium of ponderosas as a food resource, and have left evidence of this gathering activity across the mesa (Brown 1996; Wilshusen and Towner 1999).

From the 1940s to the early 1980s there was seasonal gathering of pinyon nuts and chokecherries (*Prunus virginiana*) in the park, primarily by Hispanic families, although Navajo families also used these resources. Most gathering was done near roadways, though they sometimes penetrated more deeply into wilderness areas (M. Colyer, personal communication 2003). Gathering has been discouraged in the park because of the sensitive nature of the archaeological sites, including cliff dwellings (G. San Miguel, personal communication, 6 Oct 2003). Exceptions to this rule do exist; the Hopi have permission to come to the park yearly to get water from the spring at Balcony House, and if this spring is dry, they get water from the spring at Spruce Tree House. They also gather hematite from locations within the park (G. San Miguel, personal communication 2003).

Fire History

Wildfire in the pinyon-juniper woodland at Mesa Verde tends to be stand-replacing. Mature trees, seedlings, and shallow seed banks are destroyed, and re-establishment is reliant solely on long-distance dispersal and reintroduction of seed from adjacent sites (Floyd et al. 2003b; Floyd-Hanna and Romme 1995). It may take 300 years for a severely burned mature pinyon-juniper forest to return to mature pinyon-juniper of the quality at Mesa Verde (Omi and Emrick 1980). Wildfires in the pinyon-juniper also tend to be crown fires, as these trees have low crowns, which carry the fire from the ground, resulting in higher fuel loading in live trees (Rogers 2002). Erosion is often a serious problem following large wildfires in these woodlands (Omi and Emrick 1980).

Most fires in the park are caused by lightning, although some historic fires were likely caused by humans. For example, some sources say that the Wild Horse Mesa fire of 1934 was human-caused (Meinecke 1935), as was the Park Entrance fire of 1951 (Rogers 2002). In addition, there is evidence from oral histories, park administrative documents, and historic photos (Figures 13 and 14) that the Utes deliberately used fire historically, at least in the canyon bottoms (M. Colyer, personal communication 2003; Torres-Reyes 1970:259–260). This may have been done to control encroachment of pinyon-juniper into the oak scrub–petran chaparral and riparian communities to enhance wildlife habitat and range value for cattle and horses, to aid in herding horses (Torres-Reyes 1970:259–260), and to increase the harvest of some wild plants. In a 1969 study of fire in pinyon-juniper habitats, deer and elk numbers were higher in burned areas than in adjacent unburned areas. Researchers have suggested that a series of long, narrow burns would increase game populations (Omi and Emrick 1980). Erdman (1970) estimated that 40 percent of the park was dominated by pinyon-juniper in 1970; today the estimate is around 50 percent. This increase may be due to the lack of fire (intentional or natural) in the petran chaparral habitats, which would have kept pinyons and junipers from invading. In 1969, following a study on the relationship of fire and prehistoric human occupants of

54

Figure 13. Brush burning on the Ute Mountain Ute Reservation, east of Mancos River. September 26, 1960. Photo courtesy of Mesa Verde National Park.

the mesa, Erdman et al. (1969:17–18) stated that "possibly the prehistoric Indians deliberately burned [the] upper parts of the mesa, which are of marginal farming value, in order to maintain the shrub vegetation, which supports a heavier game population than a pinyon-juniper forest approaching climax." Fire also increases many plants known ethnographically to be used by the Utes and archaeologically by the Puebloan farmers of the mesa (Adams 2002).

Since its inception in 1906, the park has had a total fire suppression policy because of concerns about damage to its more than 4000 archaeological sites. However, perhaps aggravated by 100 years of active fire suppression, or by a combination of canopy fuel buildup during two wet decades before 1995 followed by severe drought, fire frequency began to dramatically rise beginning in the mid-1930s, and there is a trend toward larger, hotter, stand-replacing wildfires (Floyd et al. 2003b, 2004; Rogers 2002). From 1926 (when fire records began) until early 1934, the park experienced no large fires (Torres-Reyes 1970), though there is some evidence

Figure 14. Range burned by Utes, Mancos Canyon, July 1933. Photo courtesy of Mesa Verde National Park.

of two large fires in the late 1880s (Erdman 1970). Since 1934, more than 80 percent of the park has burned. In the past 7 years alone, five wildfires have burned more than half of the park (Floyd et al. 2003b; Rogers 2002). Floyd et al. (2003b) concluded that a period characterized by few fires during the early twentieth century was probably caused by a combination of widespread grazing, which reduced fine fuels, and greatly improved detection and suppression of fires within the park. In the latter half of the twentieth century, however, fire patterns have been more like those observed in the late nineteenth century, though probably more frequent and more intense. According to Floyd et al. (2003b), the increase in understory vegetation following the removal of livestock from the park, and several summers of climatic conditions favorable to fire (dry and windy), together sparked the large, catastrophic fires of the latter half of the century. More recently, Floyd et al. (2004) have also shown that climate probably has played a very strong role in creating the unprecedented, intense fire regime since 1995. Though the current drought is similar in its intensity to the severe drought of the 1950s, the two decades preceding it were unusually wet, and led to a larger than normal buildup of fuels in the understory and canopy.

Floyd et al. (2003b, 2004) have devised a method of determining fire frequency in the petran chaparral zone of Mesa Verde during the last 150 years that consists of dating the successional communities that typically follow wildfire. Their conclusions indicate that fire cycles through the petran chaparral in approximately 100-year intervals. However, fire frequency in the mature pinyon-juniper woodlands tends to be much longer, on the order of about 400 years. Floyd et al. (2003b, 2004) measured fire intervals in the pinyon-juniper habitat based on analysis of stand ages, and found that many stands had trees that were at least 400 years old. They did not detect a fire frequency pattern characterized by extensive, low-intensity surface fires; instead, fires in the pinyon-juniper—both historic and modern—tended to be infrequent, but stand-replacing when they occurred (Floyd et al. 2003b, 2004).

Environmental Consequences of Cumulative History

Mesa Verde National Park has maintained a high level of biodiversity and plant endemism (Floyd and Colyer 2003). As noted above, there are several possible explanations for this. The mesa top is a natural refuge from development and grazing, and the park contains a high proportion of wilderness. It is surrounded by largely undeveloped areas, creating natural wildlife corridors and a buffer zone. Also, its rich cultural-environmental history may have created a patchwork of variously aged vegetation types within the larger landscape. A high rate of plant endemism may be due to the wide variety of substrates on the mesa, as well as the presence of the old-growth pinyon-juniper woodland, which appears to support some plants that do not grow in younger forest stands (Floyd and Colyer 2003). Because much of this biodiversity and endemism is linked to old-growth pinyon-juniper woodlands, protecting them is a high priority for resource managers. The greatest threats identified to these woodlands are climate change, fire, air pollution, and invasive exotic plant species (Romme et al. 2003a).

There is some evidence of prehistoric episodes of deforestation in the pollen record of Mesa Verde (Petersen 1994; Petersen and Mehringer 1976; Wycoff 1977) and in the archaeobotanical (macrofossil) record at nearby Grass Mesa (Kohler and Matthews 1988). Although humans could have played some role, it may be more likely that periods of drought could predispose trees to infestation by disease (such as black stain root disease) or insects (such as bark beetles), leading to large die-offs, as happened in the early 1930s and again in the mid-1970s (Omi and Emrick 1980; Floyd et al. 2003d). In addition, recent heavy fire years have been associated with especially dry periods during the normally wet July-August season (Floyd et al. 2003b). It is interesting to note that most stands in the old pinyon-juniper woodland at Mesa Verde contain trees that are about 400 years old (Floyd et al. 2003a). They would have begun growing within decades after a severe drought between 1528 and 1551 (Burns 1983). Drought could have contributed to deforestation by increasing the risk of infection or infestation by black stain root disease or bark beetles, by increasing the risk of large, high-intensity wildfires, or by a combination of these factors.

Because fire is so damaging to mature pinyon-juniper woodlands, some researchers argue that fire should not be used as a management tool in this habitat type (Floyd et al. 2004; Romme et al. 2003b, 2003c). Indeed, even thinning the dead trees in this habitat can be potentially more damaging than fire risks from standing dead trees, because of increased risks of invasive weed species introduction and removal of shade and protection for new seedlings (Romme et al. 2003b).

There are some cases, however, when fire has beneficial effects on the pinyon-juniper understory, enhancing both species richness and composition and the abundance and vigor of several plant species with potential value to humans and wildlife. Adams (2002) observed rates of recovery for burned versus unburned plots in a 5-year post-fire study after the 1989 Long Mesa fire. Many shrubs, most notably Gambel oak, began to resprout almost immediately after the fire. Many understory species also increased in frequency and cover, especially several grasses, wild tobacco (*Nicotiana attenuata*), sego lily, buckwheat (*Eriogonum* spp.), globemallow (*Sphaeralcea coccinea*), goosefoot, knotweed (*Polygonum douglasii*), and numerous other native and exotic herbs. Ethnographic uses of these taxa include basketry, for which the new straight, flexible shoots of shrubs are harvested, food (grass seeds, acorns, pot herbs, buckwheat, sego lily, etc.), and medicines, drugs, or ceremonial items (wild tobacco). Burning creates the potential for proximity to four vegetation communities (burned and unburned pinyon-juniper and petran chaparral) rather than just two (unburned pinyon-juniper and petran chaparral), with differing resources available in each community, providing a greater suite of potential resources for humans and wildlife. Fire in the petran chaparral—where fire return intervals are about 100 years (Floyd et al. 2003b)—resulted in especially quick recovery and greater biodiversity and vigor, whereas fire in pinyon-juniper woodlands resulted in greater invasion by exotic plants and slower recovery (Adams 2002). Petran chaparral also provides important habitat for mule deer (*Odocoileus hemionus*) and wild turkeys (*Meleagris gallopavo*), both important game animals (Adams and Petersen 1999).

With the catastrophic fires of the past decade have come funds to study the effects of the fires and to find ways to alleviate the risks and enhance recovery (National Park Service 2000). An important part of the recovery efforts has been to reseed the burned landscapes with native grasses. These efforts have met with variable success (Floyd et al. 2001), but park naturalists have noticed that the important cool-season mutton grass is now among the most common growing inside the park, but outside the park it is rare or nonexistent (M. Colyer, personal communication 2003). Reasons for this are difficult to pinpoint exactly, because grazing by cattle and sheep are most destructive to these cool-season bunch grasses; exclusion of livestock for 60 years may also be an important factor in the re-establishment of this grass.

It has been 70 years since native cultures have been an active (though diffuse) component of the landscape at Mesa Verde National Park, and 100 years since groups of more than a few people have lived on or used the mesa as part of their livelihood. This contrasts with the next case study, in the Chuska Mountain region and Canyon de Chelly, where Navajos still live in traditional ways on the landscape.

Hogan in Canyon del Muerto, 2003
Photo courtesy of Gary P. Nabhan

THE CHUSKA MOUNTAIN COMPLEX

The Chuska Mountain complex of the Defiance Plateau forms the only "sky island" archipelago on the Navajo Reservation. It is located just west of the Continental Divide and includes the Lukachukai Mountains and Carrizo Mountains, and Canyons de Chelly and del Muerto below the Chuska Mountains (Figure 15). This sky island archipelago extends along the Defiance Uplift, an asymmetric anticline beginning in the north near the Four Corners and Shiprock, and extending southward nearly to Interstate 40 at the Arizona–New Mexico boundary. Along its north-south axis, the Defiance Uplift extends nearly 100 miles.

There is great variation in the width of these mountain ranges on the east-to-west axis; the uplands vary from 6 miles to more than 24 miles across. The valley floors on either side of the uplift range from 6000 ft to almost 7000 ft, and the sky islands themselves rise some 2000 ft above the floor. The highest promontories in the Chuskas reach 9000 ft, sufficiently high to create a gradient of orographic precipitation; they represent the most well-watered terrain on Navajo Nation land.

Canyon de Chelly and other side canyons are shaped by streams that have cut through the Defiance Plateau and the eastern section of this uplift, the Chuska Range. Canyon de Chelly National Monument encompasses 400 sq km (Figure 16). The main drainage of Canyon de Chelly is formed by Chinle Wash and its tributaries, South Wheatfield and Whiskey Creeks. Not far from National Park Service headquarters, Chinle Wash is joined by Tsaile Creek, which drains Canyon del Muerto. Most of the exposed bedrock in these canyons is either cross-bedded de Chelly sandstones from the Permian, or Cutler red-beds. Together they form vertical cliffs in Canyon del Muerto and Canyon de Chelly that reach heights of 800–1000 ft. In addition to cliffs and slickrock surfaces of sandstones and mixed red-beds, there are slope colluvial deposits, aeolian dunes and sheets, bench and fan gravels, and mixed floodplain and valley fills occurring over large areas within this landscape. These latter geomorphic surfaces are highly erodible, and have been dramatically affected by woodcutting, grazing, farming and other cultural land management practices over the centuries (Reed and Hensler 1999).

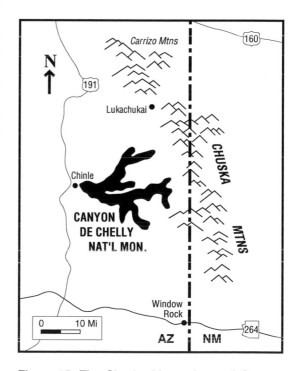

Figure 15. The Chuska Mountains and Canyon de Chelly National Monument.

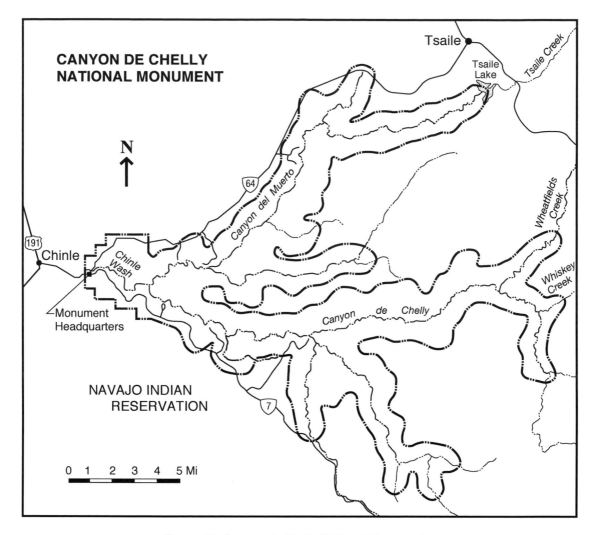

Figure 16. Canyon de Chelly National Monument.

Although the regional climate is semi-arid, there is considerable microclimatic variation within the Chuskas through space and time. Mean annual rainfall at various Navajo weather stations in this landscape varies from less than 6 inches per year to more than 11 inches per year—with a near-doubling of precipitation with elevational increases to the top of the Chuskas. The mean annual rainfall is 9.26 inches at the Canyon de Chelly weather station near NPS headquarters, situated at ca. 5540 ft in elevation. Mean monthly temperatures in the Chuskas vary from station to station by as much as 20 percent. Microenvironments within the canyon system of de Chelly and

del Muerto include alcoves, talus slopes, north-facing window boxes, and south-facing benches that juxtapose radically different microclimates within a distance of less than a third of a mile.

The Chuska Mountain complex has been dominated by ponderosa pines for most of the Holocene (Savage 1989; Wright et al. 1973). Plant fragments from packrat middens dated to 4210 BC in Canyon del Muerto suggest that most dominant plants found in contemporary vegetation in or near the canyons had arrived by that time (Betancourt and Davis 1984). Dean (1988) and others have collected tree-ring data that facilitate a relatively precise dendroclimatological reconstruction of past climates in this land-scape. In 1996 the Southwest Paleoclimate Project used four sets of tree-ring data to generate a Palmer Drought Severity Index (PDSI) analysis of climatic variation from AD 10 to 1210. Over this period, five major droughts occurred in the area (Reed and Hensler 1999). Additional tree-ring studies by Graybill and Rose (1985) and Savage (1989) suggest that more recent droughts may have occurred during the periods of 1680–1690, 1730–1750, 1800–1830, 1880–1910, and 1950–1960. At least some of these periods may have been important to vegetation change, should stand-replacing fires and/or insect infestations have occurred. As discussed in more detail below, overall wetter and cooler conditions between 1050 and 1150 appear to be correlated with a dominance of upland forest plants in the canyon pollen record, and with human population expansion in Canyon del Muerto (Fall et al. 1981).

Although Navajo botanist A. Clifford is close to completing a flora of the Chuska archipelago, no systematic inventory of landscape-level biodiversity is now available for the entire Defiance Plateau. However, Rink's (2003) flora of Canyon de Chelly and other, locality-specific checklists (such as Cove–Red Valley's by Reed and Hensler 1999) offer some insight into this remarkably diverse landscape. Until recently, the flora of Canyon de Chelly National Monument (including Canyon del Muerto) was considered to include 520 known species of vascular plants, of which about 468 taxa (90%) are natives that likely occurred there in prehistoric periods (Rink 2003). The other 10 percent are largely weeds of waste places that were introduced from the Old World. Rink's work has documented 260 previously unrecorded taxa, including 66 new genera and 12 new families in Canyon de Chelly National Monument, bringing the total to 795 taxa. Taxa accumulation curves based on random sampling efforts suggest that significant plant diversity remains to be discovered at Canyon de Chelly (Rink 2003). The canyon is now known to harbor 29 sensitive plant taxa, 22 of which are rare (Table 12).

A mammal inventory for Canyon de Chelly published by Burgess (1973) includes 41 species confirmed in the canyons themselves, with another 14 species recorded in the surrounding area of the Chuskas. Several decades of avifaunal records archived at the NPS headquarters for Canyon de Chelly National Monument also indicate considerable species richness.

Of the biotic communities found on the rims of these canyons, pinyon-juniper woodlands dominate, with big sagebrush and mutton grass common throughout. Pinyon-juniper woodlands cover almost 60 percent of Canyon de Chelly National Monument, but their flora comprise only 15 percent of the plants known in the monument (Dennis 1975). In higher and cooler locales, ponderosa pine forest dominates, with Gambel oak and big sagebrush forming sizeable stands on the edges of park-like openings (Brown 1994; Dennis 1975). Along the canyon rims, cliffs, and ledges above 6200 ft are a few species that do not generally occur in the canyons, including two yuccas, sages, one-seeded juniper, and scrub oak; above 6600 ft another 10 ecologically important perennials can be found that seldom reach into the canyon, including serviceberry, Douglas fir, and Gambel oak. Rabbitbrush, Mormon tea, and sugarberry sumac can be found on slopes and flats both within and above the canyons (Dennis 1975).

In addition to pinyon-juniper woodlands and ponderosa forests along the rim, Dennis (1975) and Harlan and Dennis (1976) have categorized the vegetation of Canyon de Chelly into five other vegetation associations related to presumed Anasazi plant resource use within the canyon: canyon bottom (riparian) communities; talus communities; springs, seeps, and other wet places; low shrub-grassland communities; and sagebrush communities.

Most plants and animals in this canyon country belong to Great Basin riparian gallery forest and wetland biotic communities. Fremont cottonwoods, willows, introduced Russian olives (*Elaeagnus angustifolia*), and tamarisks (*Tamarix* sp.) all form thickets in well-watered portions of the canyons. Hang-

Table 12. Plant species recovered from archaeological sites in the Chuska Mountain region.

Scientific Name		Common Name	Probable Uses
Agavaceae	*Yucca baccata, Y.* sp.	Banana yucca, yucca	Food, fiber
	Yucca angustissima	Narrow leaf yucca	Food, fiber
Amaranthaceae	*Amaranthus* sp.	Pigweed	Food
Anacardiaceae	*Rhus trilobata*	Skunkbush	Food
Apocynaceae	*Apocynum* sp.	Dogbane, Indian hemp	Fiber
Asclepiadaceae	*Asclepias* sp.	Milkweed	Food
Asteraceae	*Ambrosia* spp.	Ragweed	Food
	Artemisia sp.	Sagebrush	Fuel, wood, fiber, food
	Ericamera nauseosus, E. sp.	Rabbitbrush	Fuel, wood, fiber
	Helianthus sp.[1]	Sunflower	Food
	Tragopogon sp.	Goatsbeard	—
Berberidaceae	*Berberis* sp.	Barberry	—
Betulaceae	*Alnus* sp.	Alder	Fuel, wood
	Betula sp.	Birch	Fuel, wood
Brassicaceae	*Descurainia* sp.	Tansy mustard	Food
Cactaceae	*Echinocereus* sp.	Hedgehog cactus	Food
	Mammillaria sp.	Fishhook, pincushion	Food
	Opuntia spp.	Cholla, prickly pear	Food
Capparaceae	*Cleome* sp.[1]	Beeweed	Food
	Polanisia sp.	Clammyweed	Food
Chenopodiaceae	*Atriplex canescens, A.* sp.	4-wing saltbush, saltbush	Food, fuel
	Chenopodium berlandieri, C. sp.	Goosefoot	Food
	Corispermum sp.	Bugseed	Food
	Cycloloma sp.	Pigweed	Food
	Sarcobatus vermiculatus	Greasewood	Wood, fuel
	Suaeda spp.	Seepweed	Food
Convolvulaceae	*Convolvulus* sp.	Bindweed	—
Cucurbitaceae	*Cucurbita* spp.[1]	Squash	Food
	Lagenaria sp.[1]	Gourd	Container
Cupressaceae	*Juniperus* sp.	Juniper	Food, wood, fuel, fiber
Cyperaceae	*Carex* sp.	Sedge	Food, fiber
	Scirpus sp.	Bulrush	Food, fiber
Elaeagnaceae	*Shepherdia* sp.	Buffaloberry	Food
Fabaceae	*Astragalus* sp.	Milkvetch	—
	Phaseolus vulgaris, P. sp.[1]	Bean	Food
Fagaceae	*Quercus* spp.	Oak	Food, wood, fuel
Geraniaceae	*Erodium* sp.	Crane's bill	—
Hydrangeaceae	*Fendlera* sp.	Fendler bush	Wood, fuel
Hydrophyllaceae	*Phacelia* sp.	Phacelia	—
Juglandaceae	*Juglans* sp.	Walnut	Food, wood, fuel, dye
Juncaceae	*Juncus* sp.	Rush	Fiber
Lamiaceae	*Salvia* sp.	Sage	Food, medicine
Loasaceae	*Mentzelia* sp.	Stickleaf	—
Malvaceae	*Malva neglecta*	Common mallow	Food
	Sidalcea sp.	Checkermallow	Food
	Sphaeralcea sp.	Globemallow	Food
Oleaceae	*Fraxinus* sp.	Ash	Wood, fuel
Onagraceae	*Gaura* sp.	Bee blossom	—
Oxalidaceae	*Oxalis* sp.	Wood sorrel	—
Pinaceae	*Abies* sp.	Fir	Wood, fuel, pitch
	Picea spp.	Spruce	Wood, fuel, pitch
	Pinus edulis, P. sp.	Pinyon pine, pine	Wood, fuel, food, pitch
	Pseudotsuga menziesii	Douglas fir	Wood, fuel, pitch
Plantaginaceae	*Plantago* sp.	Plantain	Food
Poaceae	*Bromus* sp.	Brome	Food
	Achnatherum hymenoides	Indian ricegrass	Food
	Eragrostis sp.	Lovegrass	Food

Table 12 (continued)

Scientific Name		Common Name	Probable Uses
Poaceae (cont.)	*Phragmites australis*	Common reed	Fiber, fuel
	Sporobolus sp.	Dropseed	Food
	Stipa sp.	Needlegrass	Food
	Zea mays[1]	Maize, corn	Food, fuel
Polygonaceae	*Eriogonum* sp.	Buckwheat	Food
	Polygonum sp.	Knotweed	Food
Portulacaceae	*Portulaca* sp.	Purslane	Food
Rhamnaceae	*Ceanothus* sp.	Buckbrush	Wood, fuel
Rosaceae	*Amelanchier* sp.	Serviceberry	Wood, fuel, food
	Cercocarpus sp.	Mountain mahogany	Wood, fuel
	Prunus areniaca	Apricot	Food
	Purshia sp.	Cliffrose	Wood, fuel
	Rosa sp.	Rose	Wood, fuel, food
Rubiaceae	*Cephalanthus* sp.	Buttonbrush	
Salicaceae	*Populus* sp.	Cottonwood, poplar, aspen	Wood, fiber
	Salix sp.	Willow	Wood, fiber
Solanaceae	*Datura wrightii*	Jimson weed	Medicine, ceremony
	Nicotiana attenuata, N. sp.	Tobacco	Medicine, ceremony
	Physalis sp.	Groundcherry	Food
	Solanum sp.	Nightshade	—
Typhaceae	*Typha* sp.	Cattail	Food, fiber
Ulmaceae	*Ulmus* sp.	Elm	Wood, fuel

[1]Cultivated species.
Sources: Brandt 1998; Dean 1998; McVickar 1999; Prouty 1997; Smith 1997, 1999; Smith et al. 1999.

ing gardens and floodplain wetlands cover small areas around numerous springs and seeps. Fremont cottonwood is the most abundant tree, but most current mature trees were planted by the National Park Service to control erosion, and younger cohorts established themselves after the 1993 floods. Unfortunately, early erosion management efforts also introduced non-native tree species for the same purpose: tamarisk, Russian olive, and peach-leaf willow (*Salix amygdaloides*; Dennis 1975). These riparian thickets and wetlands are interspersed among fields, orchards, and flood-scoured flats dominated by weedy annuals. As the canyons drain into the Black Mesa basin to the west, their slopes are dominated by Plains and Great Basin grassland (shortgrass steppe) grading into desert scrub on the most arid sites (Dennis 1975).

Culture History

The Chuska Mountain complex has been experiencing "chronic human disturbance" for millennia (Savage 1989). Although abo-riginal rock art and tools from the Paleo-indian and Archaic periods have been found in the Canyon de Chelly area, human use of land, water, and vegetation began to shift dramatically around 3120 years ago, when the first evidence of incipient agriculture was recorded in the form of pollen from corn and native agrestals such as sunflower (*Helianthus* sp.; Betancourt and Davis 1984). Discussions of ancestral Puebloan (or Anasazi) occupation and land-use impacts on vegetation are typically framed in terms of the cultural sequence occurring from the Pueblo III period, through Pueblo II, Pueblo I, Basketmaker III, Basketmaker II, and a postulated Archaic period (Fall et al. 1981).

The most instructive paleoenvironmental reconstruction of Canyon del Muerto has been put forward by Fall et al. (1981) from pollen assemblages deposited between AD 700 and 1300. They noted that after agriculture began to be practiced on the canyon bottom, there was increasing aggregation and population expansion during Basketmaker III and early Pueblo I periods. Canyon bottom samples show that arboreal pollen of pines, Douglas fir, and junipers still dominated well after 700, when a demographic shift in human population began to intensify habitation and farming on the canyon bottom, accelerating the clearing of patches of forest and woodland vegetation. Between 900 and 1100, minor shifts in pollen frequencies in the record indicate an increase in shrub densities and a decrease in coniferous tree densities. Then, beginning around 1120 and intensifying through 1180, there was a dramatic shift to a dominance of shrubs on the canyon bottom, as the human population continued to expand. During this period, canyon dwellers were harvesting trees and many other resources from above the canyon rims.

Prior to AD 1100, architectural timber for the great houses of Chaco Canyon was extracted from the Chuskas, facilitated by pre-existing sociopolitical ties between Chaco and outlying communities at the base of the Chuskas (English et al. 2001). Twice as many beams found at Chaco fall in the isotopic range of living trees in the Chuska as in those of the neighboring San Mateo Mountains. Chacoans were selecting a single-age cohort of spruce and fir from both the Chuskas and San Mateos as early as 974 and as late as 1100. Between 1100 and 1250, logs from higher elevations of the Chuskas were used locally in the construction of Antelope House in Canyon del Muerto. The large structures built in these canyons used pinyon pine for roughly half of their non-viga timber—which was probably derived from the immediate surroundings (within a mile)—but the need for longer beams required that spruce and fir were transported from longer distances. In any case, there was local depletion of pinyon pines and junipers, as well as ponderosas and Douglas firs within the canyons during this time, while shrubs multiplied and expanded. There is also evidence of fuelwood depletion, to the point that poorer (but faster growing) sources of fuel such as Fremont cottonwood were cleared from the canyon bottom between 1100 and 1270 (Fall et al. 1981).

Although the unique environment of the canyon bottom has produced no evidence of ancestral Puebloan burning practices, Savage's (1989) review of tree-ring and fire scar studies from elsewhere in the Chuskas suggests that frequent low-intensity fires were common at higher elevations throughout prehistoric and historic times until the 1830s, when Navajo Churro sheep were introduced in numbers sufficient to reduce understory fuel loads and to suppress or limit the spread of fires. Monti's (2003) report on contemporary Navajo use of fire in the Chuskas suggests that in areas where fields are planted, fires are frequently set to burn crop and weed debris, but the patchwork of fields and forest thickets keeps fires from spreading too far. However, we lack data from the Chuskas to indicate whether prehistoric fire frequencies were much influenced by Native American burning practices, or were simply the result of mostly lightning-initiated, low-intensity fires.

Administrative History

Although this sky island archipelago lies almost entirely within the Navajo Reservation, influences from Puebloan (especially Zuni and Hopi), Ute, and Comanche cultures have also been recorded. As discussed in detail below, the National Park Service has also influenced the last seven decades of management of Canyon de Chelly and Canyon del Muerto, even though the canyons themselves have remained under Navajo (Diné) sovereignty and stewardship.

The Navajo Nation was formally established in 1868 in a treaty negotiated after the disastrous "Long Walk." At 17.5 million acres, it is the largest reservation in the United States. The Chuska Mountains were fully included within the original boundaries of the Navajo Nation, which has the greatest number of native-speaking farmers and foragers of any tribe north of Mexico.

The history of the establishment of Canyon de Chelly National Monument is interesting and unique within the National Park Service. In 1903, concern about systematic looting at the cliff dwellings in Canyon de Chelly prompted the appointment of Chinle trading post owner Charles Day to oversee their protection (Brugge and Wilson 1976). Unfortunately, Mr. Day appears to have been responsible for further intense looting of the ruins; he was reprimanded and left the Chinle area not long after. Archaeologists continued to voice their concerns, and the idea for establishing a national park in the area was proposed to the Navajo Tribal Council in 1925 (Brugge and Wilson 1976). The council agreed to the proposal, as long as Navajo rights to grazing and other land uses were left intact and they were allowed to supply horses to park visitors. The proposal then languished until, after some negotiation, it was again brought before the tribal council in 1930 and was accepted (Brugge and Wilson 1976). Canyon de Chelly National Monument, established in 1931, is the only such monument not owned by the federal government (Brugge and Wilson 1976). Tribal rights and family land ownership are preserved in the monument, with the National Park Service and the Navajo Nation managing the lands jointly. The NPS is responsible for the construction of all facilities and roads within the monument.

Land-Use History

The Navajo began grazing sheep in the Chuska Mountain region around 1740, about 130 years before most other places on the Colorado Plateau. The area has the longest grazing history on the reservation (Savage 1989; Underhill 1971). Grazing slowly increased in intensity, producing little noticeable change in fire frequency or pine seedling recruitment, until around 1830, the beginning of what Savage (1989) termed the Pastoral period, when grazing intensity reached a threshold that caused fire frequency to drop off sharply and permanently. The Navajo Reservation as a whole, and the Chuska Mountain region in particular, has been under "heavy and continuous grazing pressure" from the first half of the 1800s until the 1930s (Savage 1989:127). Sheep tend to browse more efficiently than cattle or native ungulates, cropping a broad range of grasses and forbs very close to the ground,

and also consuming shrub and conifer seedlings and pine needles. This reduces the fine fuels important for carrying low-intensity fires (thereby reducing fire frequency), and also negatively affects the recruitment of new ponderosas (Savage 1989). Cattle tend not to consume pine seedlings or needles, but do reduce the competition pines face from grasses and forbs in the understory, which could potentially have a positive effect for recruitment of pines. Although there were periods of wet climate favorable to regeneration of pines at several points in the 1800s, little regeneration has occurred in the Chuska Mountain complex since that time (Savage 1989). From this evidence, it appears that overgrazing began to lengthen fire intervals and generally suppress fires around 1830, without immediately creating the conditions of doghair thickets that are common in other ponderosa-dominated landscapes (Covington 2003; Savage 1989). Navajos have continued to use the woodlands and forests above the canyon rims for a variety of non-timber forest products through the present day, and these were largely derived from forest understories in the Chuskas.

Grazing intensity has fluctuated greatly over time. Spanish accounts of the late eighteenth century number Navajo herds in the hundreds of animals, but they had increased just 50 years later to number in the hundreds of thousands (Bailey and Bailey 1986; Savage 1989). Just prior to the Navajo War, their sheep numbered between 250,000 and 500,000, exceeding the carrying capacity of the range before 1850 (Bailey and Bailey 1986; Savage 1989). Upon their return to the reservation in 1868 after the Navajo War and the Long Walk, their livestock assets were 940 sheep, 1025 goats, 1550 horses, and 20 mules (Bailey and Bailey 1986). Rapid growth of the herds was the trend during the 1870s, followed by a leveling off during the 1880s. By 1882 sheep and goats totaled almost 1.72 million. Between 1893 and 1900 there was a precipitous decline in stock numbers, followed by an equally rapid recovery between 1900 and 1915 that brought total stock numbers to their highest level since the Navajo were placed on the reservation. From 1915 to 1922, Navajo horse herds were greatly reduced along with more modest declines in sheep, goats, and cattle. During the period 1923 through 1931, herds and flocks again increased but at a much slower pace; nevertheless overstocking was still very much a problem. In 1933, the Soil Conservation Service and the Bureau of Indian Affairs, judging two-thirds of the Navajo range to have been desertified by overgrazing, unilaterally instituted a severe stock reduction program on the reservation, which was greeted with little enthusiasm by Navajos. As a result of this program, stock numbers declined until well into the 1940s, and the well-adapted Navajo Churro sheep were decimated in favor of poorly adapted "modern" breeds (L. McNeal, personal communication 2002; Pynes 2000). From 1950 through 1975 there was a steady, gradual increase in animals that led again to calls for a new reduction program, a pattern that has been repeated regularly to the present day (Savage 1989).

Climate conditions favorable to pine regeneration occurred around the time of the first livestock reduction program; the release from grazing pressure and the growth of even-aged doghair thickets allowed the encroachment of pines into meadow habitats (Savage 1989). Doghair thickets in the Chuskas therefore appear to have developed two decades after those near Flagstaff. Savage noted that these structural changes were the result of "climate fluctuations ... enhanced by human alteration of disturbance regimes" (Savage 1989: 162).

As with grazing, logging also has a very long history in the Chuska Mountain region. The clearing of trees in the canyon bottoms began during Puebloan times, probably as early as AD 700, when they began to clear fields for agriculture. Increases in population densities in the canyons over time led to more clearing and harvesting of wood for fuels and construction, and between 900 and 1100 there was a shift toward more shrubs and fewer trees in the canyon bottoms (Fall et al. 1981). During construction episodes at Chaco Canyon between 1030 and 1120, an estimated 200,000 trees—including white fir, Douglas fir, Englemann spruce, and ponderosa pines—were extracted for construction at Chaco Canyon; most of these came from the Chuska Mountains and from three adjacent ranges (English et al. 2001; Betancourt and Van Devender 1983). It was not clearcutting, but extraction was of single-aged cohorts from accessible sites, a small percentage of the standing crop.

Logging was likely of little consequence on a landscape scale until 1962. Before that, although there were several small sawmills on the Fort Defiance Plateau, the Chuskas were generally considered too rugged for logging (Savage 1989). Navajo Forest Products Industries (NFPI) was established on the reservation in 1962, with an eye on the abundant old-growth ponderosa forest that blanketed the Chuska Mountains. Although ostensibly established as a benefit to the Navajo Nation—with an expected increase in jobs and payments flowing into the tribal coffers—NFPI was established with the express goal of a "planned overcut" of lumber for the first several years, and the calculated numbers for sustainable harvest were greatly overestimated (Pynes 2000). Millions of acre-feet of processed lumber were cut from the Chuskas and exported each year for the 30 years between 1962 and 1992. Although many Navajos did get jobs during this time, the costs to the landscape and to traditional uses were enormous, with long-lasting effects. Discontent among the Navajos living in proximity to the cuts is subtly evident in the tone of several of the annual reports provided by NFPI (Pynes 2000). In 1990, when logging threatened an important Navajo sacred site, Sonsela Butte, Navajo protest of this intensive logging began to heat up, and decisions made by the Navajo administration and the NFPI were called into question (Pynes 2000). Dissatisfaction deepened in the spring of 1991 when an area near Tsalie Canyon was cut severely, providing the impetus for the formation of a Navajo grassroots organization called Diné CARE to bring people together to protest the destructive overcutting. In 1993, NFPI ended their harvests in the Chuskas, and the Navajo Nation Forestry Department has released new management guidelines that include protected areas and selective cutting sites (Monti 2003; Pynes 2000).

Fire History

Savage's (1989) analysis of fire scars in the Chuska Mountains recorded fires over a period of 594 years in a 16,200-acre area. A composite fire interval over the period 1660 to 1989 was calculated to be 3.73 years, but it is the variation in fire history over time that is of most interest. This pattern shows a fairly stable fire regime of very frequent fires over 200 years until 1830. Further analysis by Savage and Swetnam (1990) indicated a fire return interval of 4.2 years for the period 1700 to 1830, but after 1830, fire scars were recorded on less than 10 percent of trees sampled, which provided evidence of only two subsequent fires in that landscape, in 1846 and 1870. This decline in the "natural" fire regime coincides remarkably with the intensification of sheep grazing on this landscape (Savage 1989). Navajo traditional uses of fire are now largely restricted to landscape mosaics in and around fields of 7–24 acres in size, but they continue at a few sites regardless of official fire suppression policies (Monti 2003).

Environmental Consequences of Cumulative History

Of the areas we studied, only the Chuska Mountain region remains a truly cultural landscape, one in which people still live in daily contact with the land and depend directly upon it for their livelihoods. It is also unique because of shared management responsibilities between the Navajo Nation and the National Park Service for part of the landscape; more than 70 families are currently involved with the NPS in a watershed restoration project. Traditional Navajo lifeways are continuing in Canyon de Chelly and throughout the larger study area, including sheep herding, small-scale farming, and gathering of plants and animals from the forests. This is not a simple wilderness area, "protected" from human impacts as much as possible; rather it is a working landscape that is dynamic in its response to humans. With changes in forest management practices after the demise of NFPI, cultural uses and values associated with forests and woodlands have begun to re-emerge as priorities. It is interesting to note that in spite of—or perhaps in part because of—the long history of human use of this region, biodiversity levels are comparable with the highest of those among our other study areas. Although a thorough floral inventory has not been completed (and work done to date suggests that there are many as-yet-unidentified plant taxa in the region), the size of the Canyon de Chelly flora may exceed even that of Bandelier National Monument, long regarded as the richest flora in the Southwest.

Table 12 lists all non-timber forest products known to have been used by prehistoric and historic dwellers of the Chuskas. Some, such as Fremont cottonwood, were used more in later prehistoric periods, perhaps due to overcutting of coniferous fuelwoods in the canyons prior to that. Others, such as beeweed (*Cleome* spp.), have apparently changed in density and in use with the introduction of agriculture and other soil disturbances. Still others, such as bugseed (*Coriospermum*), are not abundant today, but appear to have been managed and utilized intensively in prehistoric times. Many species from forests and woodlands have presumably declined in abundance with the advent of fire suppression and the development of doghair thickets. Although many of these are still used by Navajo medicine men when they can be found, the frequency of their use has declined (J. Peshlakai, personal communication 2004).

The complex interactions between grazing regimes and climate have greatly affected the ponderosa forest stand structure. This provides an interesting example of how the patterns predicted from generalizations about the interaction of grazing and stand structure may not be applicable. An understanding of the local reference envelope that shaped the ecosystems in this area is crucial to understanding the current landscape.

Grasslands at Wupatki National Monument, looking toward the San Francisco Peaks
Photo courtesy of Patrick McDonald

SAN FRANCISCO VOLCANIC FIELD

The San Francisco Volcanic Field is a chain of about 600 young volcanoes in northern Arizona, covering about 1900 square miles and including the Coconino National Forest and nearby national monuments in northern Arizona (Figure 17). These late Miocene to Holocene volcanic deposits are predominantly basalt lava flows, with some intermediate to rhyolite in composition (Holm 1986; Moore 1974; Priest et al. 2001). The San Francisco Volcanic Field also includes the only stratovolcano in the region: the San Francisco Peaks, at about 13,600 ft in elevation, were built by eruptions between 1 and 0.4 million years ago. The most recent volcano in the chain is Sunset Crater, which erupted less than 1000 years ago (Shoemaker and Champion 1977).

Wupatki and Sunset Crater National Monuments combined encompass more than 59 square miles within the San Francisco Volcanic Field (Figure 18). These monuments are located about 32.9 and 14.3 miles, respectively, northeast of Flagstaff, and are adjacent to the Coconino National Forest. The Navajo Reservation is to the east, across the Little Colorado River, and to the north and south are a mix of private, state, and BLM lands. Although these two monuments lie almost next to each other, there are important geological, historical, and ecological differences.

Sunset Crater rises from 6000 to 9000 ft. Eruption of the crater deposited black scoria tephra in the immediate vicinity and ash and cinders farther away at Wupatki (Hooten et al. 2001). Eruptions are thought to have started about 1064–1067, with intermittent eruptions probably lasting less than 50 years (Ort et al. 2002; Kirk Anderson, personal communication 2004), though some evidence suggests volcanic activity lasting up to 200 years (Shoemaker and Champion 1977).

Wupatki is, by comparison, geomorphically and topographically complex, as it is underlain by the Kaibab and Moenkopi geological formations, which are characteristic of the southern Colorado Plateau. A major fault—the Doney Fault—divides the monument in half from north to south, each side having its own geology, elevation, and dominant vegetation (Thomas et al. 2003). There is evidence of volcanic activity at Wupatki beginning 2.4 million years ago, continuing to the relatively recent eruption of Sunset Crater. The deposition of cinders and ash modified the hydrologic regime and ecotonal boundaries of the area; there is evidence throughout the park of abandoned channels, including intermittent and ephemeral streams (Blyth 1995; Travis 1990).

Wupatki and Sunset Crater are both in the rainshadow of the San Francisco Peaks. Most precipitation occurs during late sum-

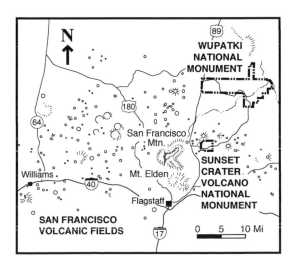

Figure 17. San Francisco Volcanic Field.

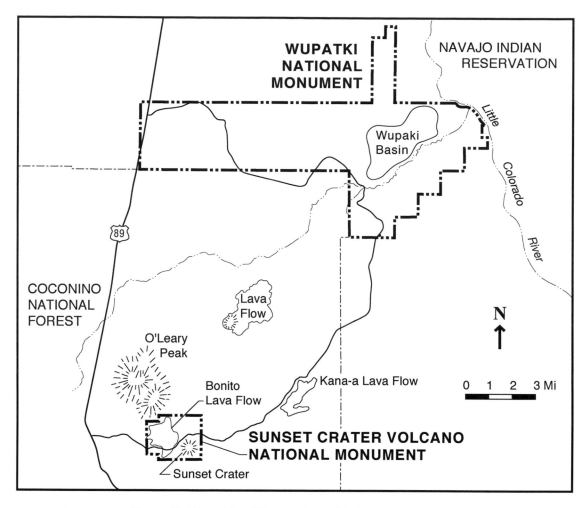

Figure 18. Wupatki and Sunset Crater National Monuments.

mer monsoon storms and in the winter, when large weather fronts originating in the Pacific Ocean are common (Anderson 1990; Colton 1960). According to the Western Regional Climate Center (www.wrcc.dri.edu/), precipitation at Wupatki averages 8.2 inches per year, and at Sunset Crater, the average total precipitation is 16.7 inches.

The San Francisco Mountains and environs are cool and semiarid and fall within the Colorado Plateau Semi-Desert province (Bailey et al. 1994). Modern vegetation at Wupatki is dominated by juniper woodland and savanna, desert scrub, and open grassland biotic communities, which are currently being encroached upon by one-seed juniper

(*Juniperus monosperma*; Cinnamon 1988; P. Whitefield, personal communication, 6 Dec 2003). The grasslands at Wupatki, which are large protected areas of native perennial bunch grasses, are considered to be in "relatively pristine condition" (National Park Service 2002b).

Conversely, Sunset Crater is characterized by a Southwest ponderosa pine forest biotic community, classified as Madrean montane conifer forest (Brown 1994). Sunset Crater's vegetation development may still be arrested or recovering from the last volcanic eruption, but it currently contains patches of pinyon pine, ponderosa pine, and aspen. Approximately 351 species of plants have been recorded within these two monuments (Anderson 1990), a flora far smaller than those of our other study areas.

Culture History

The landscape encompassing these two monuments has a rich cultural history. This area has had continuous human activity since the Clovis period, around 11,000 years ago. More than 2600 sites of archaeological significance have been recorded within the boundaries of these two monuments, and thousands more have been identified nearby outside their boundaries (National Park Service 2002a, 2002b). Archaeologists have found evidence of nomadic cultures using resources in this area dating to 11,000 and 8000 years ago, representing Paleoindian and Archaic seasonal uses. The region may have been used seasonally until almost the end of the Basketmaker III period (AD 500–675; Anderson 1993), when year-round settlement commenced with the Sinagua culture in the ninth century. Sinagua farmers, foragers, and hunters lived in the region until the temporary abandonment caused by Sunset Crater's eruptions in 1064 (Cinnamon 1988; Colton 1932). During the eruptions, it is believed that the inhabitants of the immediate crater area retreated at least as far north as the Wupatki Basin.

After cessation of eruptions, the reoccupation of Sunset Crater began with the Sinagua in the early twelfth century, and soon thereafter included the Kayenta Anasazi and Cohonina (Colton 1932; Downum and Sullivan 1990). There were marked increases in population, especially during the late Pueblo II period (ca. AD 900–1100), as archaeological evidence indicates increased complexity of farming and pit house construction in favorable soil conditions (Anderson 1993; Colton 1932; Pilles 1993; Short 1988). Large prehistoric agricultural fields were constructed in the Wupatki region—on the order of 100 acres or more. Gradual abandonment of the pueblos began around 1250, lasting nearly 500 years. The cycle of prehistoric occupation, reoccupation, and abandonment correlates with the

climatic variations around the San Francisco Peaks region at the time. Moist climatic conditions suitable for agriculture and occupation occurred during 1046–1066 and 1077–1084. The abandonment of the landscape corresponds with a period that Douglass identified as the "Great Drought," from 1276 through 1299 (Anderson 1990).

Despite this general abandonment, Hopis, Navajos, and Apaches have all had a presence on this landscape for centuries. Some Hopi clan members regard this region as their original homeland. The Hopi remain engaged with the area because it contains sacred places, capture sites for eagles, and gathering sites for medicinal and food plants. It also remains a thoroughfare for many Hopis for collecting plants from higher altitudes and more mesic habitats in the general vicinity of the San Francisco Peaks, 80 miles southwest of the reservation (Ellis 1974; Hogan and Huisinga 2000). This modern connection echoes the prehistoric habitation of the Wupatki and Sunset Crater region by the Anasazi, Cohonina, and Sinagua, who are considered by many to be the ancestors of both Hopi and Zuni people (National Park Service 2002b).

There is evidence from oral histories of both Apaches and Hopis that there were Apache bands living in this landscape as early as 1250, though archaeological evidence places them here somewhat later (C. Coder, personal communication 2004). The territory of the Mormon Lake band of the Tonto Apache included what is now Sunset Crater National Monument, and the Wupatki area was frequently traversed on trading expeditions to the Hopi mesas. The southern, eastern, and southwestern slopes of the San Francisco Peaks were important hunting grounds, and were also valuable sources for many important medicinal plants used by the Western Apache, particularly the Yu Ané (Over-the-Rim) Clan and the Mormon Lake Band (V. Randall, personal communication 2004).

This landscape is also of significance to the Navajo, or Diné people, several of whom still live within Wupatki National Monument, though others have been forcibly evicted (Hogan and Huisinga 2000; National Park Service 2002a, 2002b).

Archaeological records reveal a broad range of resources used as fuel, fiber, and food by the prehistoric people of this landscape (Table 13). Wild plants were more common foods early in the records, whereas domesticated or cultivated foods became increasingly abundant over time, indicating increased reliance upon them (Cinnamon 1988; Anderson 1990). For example, there is archaeobotanical evidence of the intensive prehistoric use of corn, cotton (*Gossypium barbadense*), various squash and pumpkins (*Cucurbita pepo*), and beans (*Phaseolus vulgaris*, *P. acutifolius*) in this area. As agriculture became more important, it is likely that "garden animals" such as rabbits (*Lepus californicus*, *Sylvilagus auduboni*), woodrats (*Neotoma* spp.), and pronghorn antelope (*Antilocarpa americana*) were also attracted to maintained fields, where they were trapped and hunted to complement vegetable food sources (Ford 1979). Anderson (1990) has suggested that in later periods, the intensity of use and the choices of these wild plant and animal resources dwindled as increasing population pressures were exerted. However, most of the animal species whose bones have been recovered during excavations remain present in Wupatki and Sunset Crater National Monuments; only a few have been extripated in the region through excessive hunting or trapping (e.g., elk and wolves; Cinnamon 1988; Schroeder 1977; Stanislawski 1963). It is interesting to note that no packrat midden dates have been recorded for the middle to late Pueblo II–III (AD 850–1250) occupation in the Wupatki area (Cinnamon 1988). Cinnamon (1988) has suggested that this could be an indication of the pressures exerted on the vegetative cover, assuming that packrat nest materials were gathered to supplement scanty fuel sources. Alternatively, prehistoric people may have hunted packrats, as they did elsewhere in the Southwest, as some indigenous people continue to do.

The Navajo began using this area seasonally in the late 1820s to hunt pronghorn antelope and to gather plant products of cultural importance such as pinyon nuts, Indian ricegrass, and Apache plume (*Fallugia paradoxa*). This landscape also served as seasonal rangelands for domestic livestock grazing (Roberts 1990). When the U.S. government signed a treaty with the Navajo in 1868 after their return from the Long Walk, this area was cited as a permanent residence for some Navajo families (Roberts 1990).

Administrative History

Wupatki National Monument was originally established on 2234 acres by President Coolidge in 1924, and in 1937 the monument was expanded to 35,254 acres; this was made possible by purchasing the adjacent land owned by Babbitt Ranches, although the ranch maintained livestock access rights through a renewable lease until 1989 (Trimble 1982). This federal designation also included limiting livestock grazing, wood cutting, and gathering of other natural resources by Navajos and Hopis. At the time of designation, Navajo residency in the monument was declared a compatible use (Hogan and Huisinga 2000), but between 1960 and 1971 many Navajos were forced out of the monument, retaining only minimal grazing rights (Roberts 1990). Sunset Crater was designated a national monument (3040 acres) in 1930 by President Hoover, at the urging of local citizens, to protect the cinder cone from its impending destruction by dynamite for use as a quarry.

Land-Use History

Sporadic livestock grazing at Wupatki may have begun with the Navajos between 1830 and 1860. From 1870 to 1898, Mormons began to settle in the Little Colorado River area, eventually reaching areas also occupied by Hopis and Navajos. These three cultural groups were soon in conflict over water rights and land access (Abruzzi 1985; Roberts 1990). The arrival of the Atlantic and Pacific Railroad in 1882 encouraged settlers and large-scale cattle ranchers, such as the Babbitt C O Bar Ranch in 1886, to move into the region from the east (Abruzzi 1995; Roberts 1990; Short 1988; Trimble 1982).

Table 13. Plant species recovered from archaeological sites in the Wupatki and Sunset Crater National Monuments.

Scientific Name		Common Name	Probable Uses
Agavaceae	*Agave parryi*, A. sp.	Agave	Food, fiber
	Yucca angustissima, *Y.* sp.	Narrowleaf yucca	Medicine, food, fiber,
	Y. baccata	Yucca	Soap
	Y. glauca	Banana yucca, soapweed	Food, fiber
Amaranthaceae	*Amaranthus* sp.	Pigweed	Food
Anacardiaceae	*Rhus trilobata*	Skunkbush	Food, fuel
Apocynaceae	*Apocynum* sp.	Dogbane, Indian hemp	Fiber
Asclepidaceae	*Asclepias latifolia*	Broadleaf milkweed	Food, fiber
Asteraceae	*Artemisia bigelovii*	Sage	Fuel, wood, fiber
	A. dracunculus	Wormwood	Fuel, wood, fiber
	A. ludoviciana, A. sp.	Sagebrush	Fuel, wood, fiber
	Brickellia californica	California brickellbush	Medicine, ceremony
	Cirsium arizonicum	Arizona thistle	—
	Ericameria nauseosa	Rabbitbrush	Dye, forage, medicine
	Erigeron sp.	Fleabane	—
	Helianthus sp.[1]	Sunflower	Food
	Pericome caudata	Mountainleaftail	Medicine
	Psilostrophe sp.	Paperflower	—
	Sanvitalia abertii	Albert's creeping zinnia	Medicine
Boraginaceae	*Cryptantha* sp.	Hollowstomach	Medicine
	Lappula occidentalis	Flatspine stickseed	—
Brassicaceae	*Lesquerella intermedia*	Mid bladderpod	Medicine
Cactaceae	*Opuntia erinacea*	Prickly pear	Food, medicine
	O. macrorhiza	Cholla	
	O. whippleii, O. sp.	Prickly pears	Food
Capparaceae	*Cleome* sp.	Beeweed	Food
Caprifoliaceae	*Sambucus* sp.	Elderberry	Medicine
	Symphoricarpos sp.	Snowberry	Medicine
Chenopodiaceae	*Atriplex canescens*, A. sp.	Four-wing saltbush, saltbush	Food, fuel
	Chenopodium berlandeiri, C. sp.	Goosefoot	Food
	Corispermum americanum	American bugseed	—
	Cycloloma atriplicifolium	Winged pigweed	Food, dye, medicine
	Krascheninnikovia lanata	Winterfat	Forage, medicine
	Sarcobatus vermiculatus, S. sp.	Greasewood	Wood, fuel
	Suaeda sp.	Seepweed	—
Convolvulaceae	*Evolvulus nuttallianus*	Shaggy dwarf morning glory	—
Cucurbitaceae	*Cucurbita pepo*, *C. mixta*[1]	Pumpkin, squash	Food
	C. foetidissima	Buffalo gourd	Food, soap
	Lagenaria siceraria[1]	Gourd	Container
Cupressaceae	*Juniperus monosperma*, *J.* sp.	1-seed juniper, junipers	Food, wood, fuel, fiber
	J. scopulorum	Rocky Mountain juniper	Food, wood, fuel, fiber
Elaeagnaceae	*Shepherdia rotundifolia*	Roundleaf buffaloberry	Food, fuel
Ephedraceae	*Ephedra viridis*, *E.* sp.	Mormon tea	Medicine
Fabaceae	*Astragalus* sp.	Milkvetch	Medicine, ceremony
	Phaseolus acutifolius, *P.* sp.[1]	Tepary bean, beans	Food
	P. lunatus, *P. vulgaris*[1]	Bean	Food
Fagaceae	*Quercus pungens*, *Q.* sp.	Oak	Food, wood, fuel

Table 13 (continued)

Scientific Name		Common Name	Probable Uses
Juglandaceae	*Juglans* sp.	Walnut	Food, wood, fuel, dye
Juncaceae	*Juncus* sp.	Rush	Fiber
Loasaceae	*Mentzelia* sp.	Stickleaf	—
Malvaceae	*Gossypium hirsutum, G.* sp.	Cotton	Food, fiber
	Sphaeralcea sp.	Globemallow	Food
Nyctaginaceae	*Mirabilis multiflora*	Four o'clock	—
Pedaliaceae	*Proboscidea louisianica, P.* sp.	Unicorn-plant, ram's horn	Ceremony
Pinaceae	*Pinus edulis*	Pinyon pine	Wood, fuel, food, pitch
	P. ponderosa	Ponderosa pine	Wood, fuel, food, pitch
	Pseudotsuga menziesii	Douglas fir	Wood, fuel, pitch
Poaceae	*Achnatherum hymenoides*	Indian ricegrass	Food
	Andropogon hallii	Sand bluestem	—
	Aristida purpurea	Fendler's threeawn	—
	Bouteloua eriopoda	Black grama	—
	B. gracilis	Blue grama	Food, fiber
	Bromus sp.	Brome grass	Food
	Elymus elymoides	Squirreltail	—
	Enneapogon desvauxii	Nineawn pappusgrass	—
	Hesperostipa comata	Needle and thread	Tool
	Hordeum jubatum	Foxtail barley	Food
	Muhlenbergia porteri	Bush muhly	—
	Phragmites australis	Common reed	Fiber, fuel, smoking
	Setaria leucopila	Streambed bristlegrass	—
	Sporobolus sp.	Dropseed	Fuel, textile
	Zea mays[1]	Maize, corn	Food, fuel
Portulacaceae	*Portulaca oleracea, P.* sp.	Purslane	Food
Pteridiaceae	*Cheilanthes feei*	Slender lipfern	—
Rosaceae	*Amelanchier* sp.	Serviceberry	Wood, fuel, food
	Cercocarpus montanus	Mountain mahogany	Wood, fuel, dye
	Fallugia paradoxa	Apache plume	Medicine, fuel, crafts
	Purshia mexicana	Cliffrose	Wood, fuel
Salicaceae	*Populus fremontii*	Cottonwood, poplar, aspen	Wood, fiber
	Populus spp.		
	Salix spp.	Willow	Wood, fiber
Scrophulariaceae	*Cordylanthus* sp.	Birds' beak	—
Solanaceae	*Datura wrightii*	Jimson weed	Medicine, ceremony
	Lycium pallidum	Wolfberry	Food, wood, fuel
	Nicotiana attenuata, N. sp.	Tobacco	Medicine, ceremony
	Physalis sp.	Groundcherry	Food
	Solanum jamesii	Wild potato	Food
Ulmaceae	*Ulmus* sp.	Elm	Wood, fuel
Vitaceae	*Vitis arizonica*	Canyon grape	Food, medicine
Zygophyllaceae	*Kallstroemia californica, K.* sp.	Caltrop	Medicine

[1]Cultivated species.
Sources: Anderson 1990; Berlin et al. 1990; Cinnamon 1988; Smith 2004; Stanislowski 1963.

Domestic livestock production occurred on land originally commissioned by Congress to the railroad company, which was later sold to private interests; however, this was land upon which Navajos were already established. Initially, 1 million acres were sold to the Aztec Land and Cattle Company at a price of only 50 cents per acre, in 1884. These land transactions occurred without any Navajo involvement (Roberts 1990). Later, an additional 1 million acres from the Atlantic and Pacific Railroad Company were acquired by other private interests, which triggered the importation of more than 60,000 additional head of cattle to northern Arizona around 1887. The estimated number of cattle grazing in Fort Valley, one of the San Francisco Peaks summer ranges, was 15,000 (Wallace 1949).

Acrimonious disputes between Anglo ranchers of the C O Bar and Circle S and the long-established Navajo herdsmen forced the Navajos out of the area in the 1890s. They were ordered to cross the Little Colorado River in 1897 and to remain on the east side of the river, far from the mountains (Roberts 1990). Thousands of dollars worth of Navajo sheep were lost during this crossing (U.S. Commissioner of Indian Affairs 1897, cited in Roberts 1990). In 1908, a legal battle eventually returned grazing allotments in the Wupatki Basin to the Navajos.

The ecological impacts of livestock on the Little Colorado River basin grasslands were huge, accelerating erosion and soil compaction and reducing forage quantity and quality, especially of Indian ricegrass (P. Hogan, personal communication 2001). Grazing also contributed to shifts in the dominant plant species and cover, invasion by exotic plants, altered vegetation structure (Abruzzi 1995), arroyo cutting, accelerated soil erosion, and alteration of the hydrologic regime (McFadden and McAuliffe 1997; National Park Service 2002b). Erosion, compounded with crop failures, led to abandonment of the remaining Navajo agricultural fields during the major drought of 1892–1905. During this period, thousands of livestock died because of a lack of forage (Abruzzi 1989; Roberts 1990). Before the "Dust Bowl" drought of

the 1930s, Navajo livestock populations had become quite large (up to 1,300,000 sheep). The environmental degradation caused by the combination of sheep and drought prompted the federal government to initiate the most drastic livestock reductions anywhere in the United States. The program called for slaughtering nearly 70 percent of Navajo sheep, especially Churro sheep (National Park Service 2002b; Roberts 1990). This disrupted traditional Navajo grazing patterns on a regional scale and irrevocably changed their culture and livelihood.

However, light livestock grazing continued on the Babbitt ranches and in some parts of Wupatki, as agreed upon by renewed leases between the National Park Service and the Babbitts, until it was finally phased out at Wupatki in 1989 (Abruzzi 1995; Cinnamon 1981). One Navajo family still retains the grazing rights for 40 sheep (National Park Service 2002a; P. Whitefield, personal communication 6 Dec 2003).

Many non-timber forest products are important to the indigenous cultures of this landscape, in the past and to the present day (Hogan and Huisinga 2000). More than half of the 65 plant genera found in the archaeobotanical record are known to still occur in the monuments today (see Table 13). At least seven of these (*Astragalus*, *Datura*, *Phragmites australis*, *Pseudotsuga menzeisii*, *Salix*, *Sphaeralcea*, and *Sporobolus*), have known or potential ethnobotanical value. Some taxa found in the archaeobotanical record that are now absent or that have been observed just a few times on this landscape include pungent oak (*Quercus pungens*), Rocky Mountain juniper (*Juniperus scopulorum*), willows, roundleaf buffaloberry (*Shepherdia rotundifolia*), canyon grape (*Vitis arizonica*), banana yucca (*Yucca baccata*), and soapweed yucca (*Y. angustissima*).

Scientific uncertainty about the effects of human foraging may have influenced National Park Service policy, which has periodically restricted the harvesting of non-timber forest products in these two national monuments. At a recent meeting with NPS resource managers, the Hopi Cultural Preservation Office staff and former Navajo residents of the monuments laid down a litany of examples of traditional harvesting being disrupted by NPS staff, including elderly Hopi women being "busted" for harvesting small quantities of annual rosemary-mint (*Poliomintha incana*; L. Kuwanwishma and J. Peshlakai, personal communication 2004). There are currently conflicts of interest between the National Park Service mission statement and the American Indian Religious Freedom Act (AIRFA) with respect to the gathering of ceremonially important plants and animals that are listed as threatened or endangered. However, the ancestral ties to the resources within the monument are recognized and there are some efforts to sustain this relationship through developing fair Special Use permits (Anderson 1990; P. Whitefield, personal communication 2004). Certainly, the problems of drought and wildfire also threaten the persistence of culturally utilized plant populations, and should be considered in current and future management plans.

Fire History

There is little reliable information regarding fire frequencies in both pinyon-juniper woodlands and grasslands on the Wuptaki–Sunset Crater landscape prior to 1985. There is no direct tangible evidence for historic or prehistoric burning by Native Americans in these areas. The pre-European fire frequencies for the Colorado Plateau pinyon-juniper system are still being debated. In contrast to the Mesa Verde region, the historic and natural fire regime of the pinyon-juniper woodlands in Walnut Canyon National Monument is believed to have fire return intervals of every 15 to 30 years (Despain and Mosley 1990). Although the role of fire is considered to be important for maintaining shrub steppe communities, it is also thought that fire may have kept pinyons and junipers restricted to rocky outcroppings where there was insufficient fuel in the understory to threaten their survival (Cinnamon 1988; Miller and Tausch 2001; Wright et al. 1979). Prescribed burns have successfully eradicated junipers less than 4 ft tall, but

these burns did not kill trees more than 6.5 ft tall (Bunting 1986; Cinnamon 1988).

Fire frequencies are unknown in grasslands, but grazing probably reduced fire frequency because fine fuels were not as available (Bunting 1986). Since the removal of most grazing at Wupatki, fires of small spatial scale have dotted the landscape—four since 1995. The most recent was the Antelope Fire of 2002, which scorched the southwestern end of the monument and attracted vertebrate herbivores to the vegetation on the burned sites where plant cover and vigor increased after the fire (P. Whitefield, personal communication 2004).

In contrast, dendrochronological records from ponderosa pine forests near Flagstaff demonstrate fire frequencies of 4.4 to 17 years prior to actively mandated fire suppression in 1912 and alteration of the forest structure (Dieterich 1980; Friederici 2003b). However, it is likely that fire occurring on volcanic soils can further aggravate already water-stressed trees and increase the probability of ponderosa mortality (Fulé et al. 2002). Nevertheless, NPS staff members at Wupatki and Sunset Crater believe that fire is necessary to bring greater biodiversity back into the monuments; a fire effects monitoring program was established in 2002 to implement that approach (National Park Service 2002a, 2002b).

Environmental Consequences of Cumulative History

The vegetation mosaic existing today in the Wupatki and Sunset Crater landscape has developed in response to particular prehistoric and historic land uses, such as grazing, hunting, gathering, and fire suppression, but it is also influenced by precipitation dynamics and the complexity of the geomorphic landscape (Cinnamon 1981, 1988; P. Whitefield, personal communication 2004). Initial fire suppression was probably the de facto result of grazing, which also encouraged juniper encroachment into grasslands; the national fire suppression policy enacted in 1912 reinforced this trend. Climatic conditions between 1912 and 1918 were suitable for juniper, pinyon, and ponderosa seed germination. Large cohorts were recruited, later forming doghair thickets of pines (Covington and Moore 1994; Covington 2003).

An ecotonal shift in the post-settlement period occurred simultaneously with the onset of heavy grazing and a more mesic climate in the early 1910s. Junipers expanded from the rocky outcroppings into deeper, well-drained soils, where the herbaceous understory had already been removed by grazing; previously, it could have supplied sufficient fuel to carry fires, effectively preventing juniper establishment. This change in vegetation dynamics has resulted in a longer fire return interval. The current drought is further stimulating juniper dominance and ecotonal boundary shifts as junipers are much more drought tolerant and not as susceptible to beetles as the pinyons with which they often compete. Due to the encroaching junipers, grassland habitat for pronghorn antelope is decreasing, and the remaining grasslands in and around Wupatki and Sunset Crater have become increasingly important for the local herd (National Park Service 2002b). The juniper expansion into the Wupatki grasslands is undoubtedly linked to a corresponding loss of biodiversity in the grasslands (Jackson et al. 2002).

These changes also favor the establishment of non-native species, such as red brome (*Bromus rubens*), tamarisk, rabbitbrush, threeawn grass (*Aristida* spp.), and Russian thistle (*Salsola pentandra*), which threaten native plant survival (Cinnamon 1988; National Park Service 2002b; P. Whitefield, personal communication 6 Dec 2003). Although native grasses such as Indian ricegrass, blue grama, and bluestem persist in the grazed grasslands of Wupatki, they have declined in overall abundance. Extensive patches of snakeweed and other disturbance plants have proliferated in high-use areas like trails and picnic areas. Nevertheless, because grazing is no longer permitted in the monuments, several culturally important plant species are found in much greater

abundance within the monuments than outside (National Park Service 2002b).

To provide insights into possible global climate change scenarios, the Merriam-Powell Center for Environmental Research is examining ecological responses to the stressful conditions resulting from the Sunset Crater eruptions that occurred about 900 years ago. They are studying the performance of pinyon pines on cinder soils (high-stress sites) and sandy-loam soils (low-stress sites). Sandy-loam soils are more prevalent at Wupatki, whereas Sunset Crater has cinder soils. Sunset Crater is also dominated by nutrient-deficient soils (Cobb et al. 1997; Selmants et al. 2003) that tend to drain quickly and therefore experience greater soil-water deficiency. As a result of these nutrient and water limitations on cinder soils, pinyons grow slower, produce fewer cones, and are highly susceptible to two major insect herbivores—a stem-boring moth and a sap-sucking pinyon-needle scale (Mopper et al. 1991a). The impacts of these herbivores are dramatic, impacting the entire community of dependent species more on the cinder soils than on the sandy loams.

Gehring and Whitham (1994) have shown that ectomycorrhizal associations with pinyons are more abundant on high-stress sites and assist with nutrient availability, allowing populations to tolerate these stressful conditions. Additionally, Gehring et al. (1998) showed that the ectomycorrhizal communities differ between the soil types in taxonomic patterns and relative abundance.

Other soil-driven effects have been documented to extend not only to mycorrhizal associations, but to other species such as understory plants, birds, rodents, and insects (Brown et al. 2001; Kuske et al. 2003; Ogle et al. 2000; Ruel and Whitham 2002; Swaty et al., in press; Trotter et al. 2004). For example, the abundance of arthropods and species richness is higher on trees at low-stress (sandy-loam) sites (Trotter et al. 2004). However, there are squat, shrubby pinyons in the high-stress (cinder) sites that are more susceptible to herbivory, which greatly reduces vigor (Cobb et al. 1997, 2001; Mopper et al. 1991b; Trotter et al. 2002; Whitham and Mopper 1985). Currently, the Merriam-Powell Center's researchers suggest that the current drought will devastate local populations of the approximately 1000 species associated with pinyon pines on high-strees (cinder) sites (N. Cobb, personal communication 2003).

Even within the same landscape, ecological responses of pinyon-juniper associations on volcanic cinder soils are dramatically different from those on limestone-derived loam soils, with a cascade of effects evident in the ectomycorrhizal, understory plant, insect, and bird communities. Thus, even in the same San Francisco Volcanic Field landscape, just the shift in soil substrates makes it difficult to develop a single management plan for pinyon-juniper woodlands that applies equally well to habitats on limestone-derived versus volcanic cinder soils. In this system, understanding the interactive and often cascading influences among biotic and abiotic factors, such as temperature, precipitation, genetic diversity, herbivory, ectomycorrhizal associations, and stand age, will be crucial for developing site-specific management strategies (Brown et al. 2001; Ogle et al. 2000; Ruel and Whitham 2002; Swaty et al., in press). It is thus clear that prescriptions for restoration developed for ponderosa forests in the Flagstaff area (Covington and Moore 1994) have little relevance to the diverse pinyon-juniper woodlands occurring in the same volcanic field.

The examples throughout this case study exemplify the complexity of confounding factors and the influence they have upon each other. They clearly do not lend themselves to a one-size-fits-all prescription for woodland restoration at Wupatki and Sunset Crater. It is important to consider treatments specific to the particular vegetation types or particular soils in light of how they have been previously shaped by cultural land uses.

SUMMARY OF CASE STUDIES

Our four study areas exhibit differences in physical factors, such as substrate, bedrock geology, microclimate, and dominant vegetation types, as well as differences in cultural influences, and each area faces different threats in the future. Mesa Verde and the Chuska Mountains have a predominantly sedimentary substrate, whereas the Jemez Mountains and Wupatki are primarily volcanic. Mesa Verde and the Chuska Mountains are mesic, the Jemez Mountains are moderately dry, and Wupatki is exceedingly dry. The Chuska Mountains are dominated by ponderosa pines and the Jemez Mountains are co-dominated by ponderosas and pinyon-juniper, and both mountain ranges contain higher elevation mixed conifer forests. Mesa Verde is dominated by pinyon-juniper and Wupatki contains a grassland with invading pinyon pines and juniper located adjacent to a ponderosa pine forest.

Cultural influences also vary widely. Canyon de Chelly and the Chuska Mountains have had long occupation by Navajos, and are still occupied by Navajos who continue with their long traditions of herding, farming, and gathering. The Jemez Mountains represent one of the richest cultural regions in the Southwest, with 14 long-established pueblos in the area, and with Hispanic people living in the region since 1590. Farming has been an important shaper of this landscape for at least 3000 years. Wupatki has been only lightly occupied since around 1300, but before the 1300s there were fields larger than 100 acres that were dry-farmed, indicating that this prehistoric land use was probably more intensive than any in historic periods. Both Hopi and Navajo clans still gather plants and animals there. Mesa Verde has had a very light human occupation since 1300, but before that time it was occupied rather intensively by Puebloan farmers, who probably had a strong localized influence on the forest through fuelwood and construction beam cutting, and clearing of agricultural fields in areas near habitation sites.

Livestock grazing has had a profound impact across the West, and its influence on our study areas has probably been underestimated by land managers. At Jemez, early and continuous grazing of sheep, along with some cattle and horses, influenced the landscape from the 1600s to the 1930s. In the Chuska Mountains, grazing of primarily sheep started early, around 1740, with periods of overstocking, and it still continues to this day. Mesa Verde had a relatively short grazing history, primarily of cattle, limited to about 50–60 years. At Wupatki, grazing by Navajo and Mormon ranchers began between 1830 and 1860, followed by inappropriately high stocking rates from 1870 through 1930, with very little grazing continuing to the present.

Logging has also had different effects on our study areas, with perhaps the most profound being in the ponderosa forests of the Jemez and Chuska Mountains. However, woodcutting for fuel continues on all landscapes.

These studies have provided the opportunity to see how some specific understory taxa have been influenced by land-use history. At Mesa Verde, studies documenting dramatic changes in understory plant species composition after fire indicate that some plant resources important to indigenous people used to be much more abundant, particularly tobacco and several food and basketry resources. At Mesa Verde and also in the Jemez Mountains, the effects of livestock on native cool-season grasses were detrimental. Native cool-season and bunch grasses have been shown to be important in controlling erosion, which has been an important management issue at Bandelier National Monument. Also at Mesa Verde, riparian and canyon-bottom vegetation has shown dramatic recovery after the exclusion of livestock.

These four study areas are now facing different threats (Table 14). Wildfire has not been identified as a major threat in all four landscapes, and its effects in woodlands versus forests are quite different. Habitat loss and fragmentation and invasive species are identified as major threats in three of the four study areas. At Mesa Verde, the Jemez Mountains, and Canyon de Chelly, 10–17 percent of the floras are introduced species. In addition, climate change threatens to produce vegetation shifts across the entire region, and is a factor that needs to be included in all future management plans for these study areas.

Table 14. Current threats in our study areas, as identified by land managers.

Jemez Mountains
    Urban growth and habitat fragmentation
    Wildfire
    Erosion
    Invasive plant species

Mesa Verde
    Wildfire
    Invasive plant species
    Air pollution

Chuska Mountains
    Invasive plant species
    Rural population growth and habitat fragmentation

Wupatki and Sunset Crater National Monuments
    Juniper encroachment into grasslands
    Rural population growth and habitat fragmentation

Fish Creek. Photo courtesy of Patrick McDonald.

Threats to the woodlands and forests of the Colorado Plateau include some of the same historic players described previously (logging, drought, disease, grazing, fire), and some new ones as well. Increasing development and human occupation and use of wildlands on the Colorado Plateau have brought with them the threats of habitat fragmentation and loss, pollution, a lower water table, loss of native biodiversity, and invasive plant species. Management of public lands, which has for a century excluded or discouraged traditional uses by Native Americans, has also led to loss of traditional ecological knowledge and use, which was once a significant driver of microhabitat diversity and structural dynamics. Here we describe some of these new threats and their implications for management and restoration.

Global Climate Change

Carbon dioxide in the earth's atmosphere has increased by more than 30 percent since 1750. There is broad consensus in the scientific community that human-caused emissions of carbon dioxide and other gases are making a measurable contribution to global warming (IPCC 2001; Union of Concerned Scientists 2004). Modern carbon dioxide values exceed exponentially the high-resolution paleo carbon dioxide profiles from the last approximately 400,000 years (Barnola et al. 2003). Although there is no precise date for how quickly atmospheric carbon dioxide will increase, the IPCC gives estimates ranging from doubling in 2100 to tripling or beyond by that time, depending on the level of concerted global mediation (IPCC 2001;

Schneider 2002). The modern level is above 300 ppm and rising (Neftel et al. 1994). With a doubling of carbon dioxide, Flannigan et al. (1998) predict that in Canada, the annual area burned would increase by 50 percent, and the number of lightning-caused fires is predicted to increase by 30 percent (Price and Rind 1994). The greenhouse effect from anthropogenic gases has apparently increased the earth's temperature by about 0.6° C (1.08° F) since the 1850s. Three of the 10 warmest years in the historic period of weather records have occurred since 1990, with the maximum temperature documented in 1998. Some climatologists believe that temperatures could have been even higher, but for the cooling effect of aerosols from the 1991 Mt. Pinatubo eruption in the Philippines. Dr. Webb of the USGS has recently stated that for the Colorado River watershed, the current drought is more severe than that of the Dust Bowl in the 1930s (Wagner 2004).

The anthropogenic effects on climate are not restricted to global-scale greenhouse warming. Researchers now also recognize significant local impacts on temperature and precipitation patterns from land uses such as irrigation, pollution, overgrazing, and urban development (Couzin 1999). For example, Mexico is heating up at the United States–Mexico border from the severe overgrazing that has decimated vegetation and the cooling effect from transpiring plants. On a hot day, the temperature can differ by as much as 4° C (7.2° F) within 50 ft from the U.S. side to the Mexico side (Couzin 1999). Urban heat island effects occur around Phoenix, Las Vegas, and Albuquerque, all near the

Colorado Plateau. The national fire plan is focused on suppressing severe wildland fires and reducing fuels, but there is no provision for understanding and working with the synergy of fires and climate change.

Climate over the next few hundred years will undoubtedly be substantially different from the past 150 years, and not merely a matter of degree. High-resolution paleoclimate studies of the last 10,000 years have pinpointed floods, droughts, and cold periods more extreme than any historic climate event (Easterling et al. 2000; Houghton et al. 2001; Overpeck 1996). Based on paleoclimate records, Tausch et al. (2004) have predicted that the western United States is at the beginning of a warm, dry period that could last for 100–300 years. Future weather trends forecast by complex computer simulations (general circulation models, GCMs) include higher minimum temperatures, fewer frost days, extreme precipitation events with associated floods and erosion events, and increased summer drying and drought (IPCC 2001). Average annual temperatures may increase by as much as 4° C (7.2° F) over the next 50 years, which is the same scale of temperature change that occurred at the glacial to interglacial transitions.

Major vegetation changes are sure to follow a global temperature change of several degrees. The average annual temperature in the Southwest has warmed by about 1° C (1.8° F) since 1940 (Lynch and Swetnam 2003), and between 1998 and 2003, there have been extreme summer droughts. The warmer winters and the summer droughts have favored infestations of insects that have killed water-stressed trees over large tracts of land (Anderson et al. 2003b; Anhold et al. 2001; Romme et al. 2003b). The result is a 70–100 percent die-off in many Southwest pine-dominated forests and woodlands (Anderson et al. 2003b; Anhold et al. 2001), which is more severe than the tree mortality documented in the legendary 1950s New Mexico drought (Allen et al. 2003).

United States forests are key resources in national political policy formulated to reduce global warming (IPCC 1998, 2000). The U.S. Climate Change Action Plan in 1993 pledged to reduce greenhouse gases to 1990 levels by the year 2000. Various carbon budget assessments and models (e.g. Birdsey et al. 2000) have focused on evaluating the capacity of U.S. forests to store carbon, and on reducing industrial emissions through more energy-efficient systems. Catastrophic wildfires, such as have occurred in the last 5 years, forest mortality from insect outbreaks, erosion, and other land-use impacts can potentially cause large, rapid losses of global carbon stocks (Breshears and Allen 2002). Thus at the same time that U.S. land stewards are planning for restoration, thinning, and controlled burning, there are international political referendums mandating that these same stewards limit logging and suppress fires. There is a dangerous "catch 22" in all of this as continued buildup of fuels will catastrophically subtract carbon stocks as large tracts of land burn, which could accelerate global warming processes in a positive feedback loop.

Forest restoration programs rarely incorporate discussions of the historic and future climate in their management plans, and may misinterpret appropriate reference conditions because the cause and effect links of climate are missing (Millar and Woolfenden 1999). The period before Euro-American settlement is the commonly used benchmark for restoration, but this was a distinct cool and wet period. Millar and Woolfenden (1999) stated that "restoration of Little Ice Age conditions (i.e. presettlement forests) makes little sense for the current climate period." The whole package of future climate change drivers—the current natural drought cycle, the predicted effects of global warming, and local to regional land-use impacts—will interact and potentially amplify warming trends. It seems likely that there will be extreme climate events in the future (Easterling et al. 2000; National Research Council 2002). It is clear that there is no going back to presettlement conditions—even if we could discern them adequately—and restoration efforts need to plan for future climate trends.

Traditional Ecological Knowledge

The traditional ecological knowledge of indigenous people has produced subtle cultural influences on the landscape; the use of fire is not the only such influence. Across the pine-dominated ecosystems of the Colorado Plateau, more than 900 understory species have been harvested as traditional non-timber forest products by Native Americans and Hispanic Americans for use as foods, medicines, and craft or utilitarian items, or for their ceremonial value (Anderson 1990; Berlin et al. 1990; Brandt 1999; Colyer 2003; Dean 1999; Foxx 2002; Foxx and Tierney 1985; Jacobs 1989; Kohler and Matthews 1988; Moerman 1998; Toll and McBride 1996). In the Grand Canyon wildlands ecoregion, many of these species are still used by Navajos and Hopis. They are not merely "collected," as if Native American wildcrafters passively accept what is placed before them. Instead, various species of these plants must be pruned or burned to render their products useful; others must be dug up and their corms or rhizomes separated and replanted to allow the population to persist. In the case of three-leaf sumac, old, tough, jagged branches are virtually useless to basketmakers (Anderson 2000). But when, as Adams (2002) and Bohrer (1979) have documented, new velvety shoots rise from the ashes in the month following a summer fire, these shoots are straight, long, and pliable—ideal for basket weaving. Anderson (1991) and Nabhan and Anderson (1991) have documented similar morphological changes with redbud, mountain mahoghany, and other shrubs; the utility of every plant is not a given, but comes from traditional ecological knowledge on when and how to burn or prune.

This category of natural resources is valuable and of concern, not only locally but globally, as demand increases for wild harvested herbs, medicines, and other NTFPs (Everett 1997). Only within the last 15 years have NTFPs been recognized as a category that needs to be addressed in U.S. forest management practices (Davidson-Hunt et al. 2001; Jones et al. 2002, 2004; USDA 2001).

Less than 2 percent of forest management plans cover NTFPs, and national legislation does not mandate that non-timber forest products be addressed in forest planning (Jones et al. 2002). The NPS policies on NTFP harvesting in national parks and monuments have increasingly judged Native American harvests to be sustainable until proven otherwise, but this policy shift is not realized by many rangers, who still discourage harvesters with legitimate rights to plants in parks (Hansen 2002). This is of concern because goods produced from native plants have become an active expression of cultural survival and conservation of indigenous knowledge (FAR 2003). The loss of traditional ecological knowledge that has resulted from a century of exclusion from public lands has reached a critical stage. The last two generations have lost native languages and traditional cultural education; in many cases, much of this knowledge is now preserved in books. Among the Western Apache, for example, prior to World War II, 11-year-old children were expected to be able to identify and describe dozens of native plants and their uses; today, they can only identify a handful of common plants, and often have no understanding of their uses or past importance in Apache culture (V. Randall, personal communication 2004). Further exacerbating this loss of knowledge about traditional uses of native plants is the fact that many native foods, once lost from common consumption, are unpalatable to younger people who have not grown up with them, such as the loss of palatability of acorns among the Yavapai-Apache (Coder et al. 2004). Habitat loss and destruction of former gathering grounds has also reduced the Apaches' ability to utilize formerly common and important natural resources (Pilsk and Cassa 2004).

There is now some recognition of the importance of NTFPs both culturally and ecologically (Jones et al. 2004; USDA 2001). Yet, the lack of information regarding what plants are being harvested, and from where, adds to the complexity of managing for them. In the absence of a complete scientific understanding of many of the individual species and communities affected by harvesting, the intimate knowledge gained over years of harvesting from an area can be best provided by the harvesters themselves (Coder et al. 2004; Jones et al. 2004), who are familiar with plant abundance, status, and habitat specifications. Jones et al. (2004) have noted that the level of trust and communication between harvesters and forest managers is poor, however, with biases on both sides, and harvesters do not readily disclose information about gathering sites, especially for medicinal plants. It is to varying degrees accepted and recognized as necessary for indigenous people to harvest items under the protection of the American Indian Religious Freedom Act, and there are currently a few collaborative efforts between these stakeholders to manage for and understand the status of NTFPs (Jones et al. 2004).

It appears that as many as 65 non-timber forest species traditionally used by the Navajo and Apache have diminished in abundance as a result of forest cover changes and fire management and grazing policy over the last century (Lisa Classen, personal communication 2003). Some of these species persist in harvestable quantities only in or on the edges of open patches, near fields, and in house yards in the forest that are maintained by humans as open habitats. There—despite fire suppression policies and gathering restrictions irregularly enforced by federal and tribal agencies—indigenous families and clans are maintaining a patchwork of habitats that meet their various material and spiritual needs for foods, fibers, medicines, and ceremonial paraphernalia.

It is worth pointing out that some of the habitats that give parks their very character might look, feel, and smell radically different if they had not been influenced for centuries by Native American farmers and foragers.

Invasive Plant Species

After habitat fragmentation and loss, invasive exotic species pose the second most urgent threat to biodiversity worldwide (Baker 2001). Invasive species can have far-reaching and long-lasting effects on local ecosystems, with the potential to cause radical changes in the abundance of native species, either by hybridizing with them or by out-competing with natives for scarce resources—sometimes leading to extinction of the native species (Ellstrand 2003). Roughly 400 of the 958 species listed as threatened or endangered under the Endangered Species Act are at risk primarily because of competition with or predation by exotic species (Nature Conservancy 1996; Wilcove et al. 1998). They can also alter ecosystem processes, such as siltation and flow rates in riparian habitats (Zavaleta 2000), nitrogen cycling (Vitousek 1990), or fire frequency (D'Antonio and Vitousek 1992).

There are economic consequences of invasions, as well. Not only are crop pests a growing problem, costing U.S. agriculture about $26.4 billion per year (Pimentel et al. 2000), but losses are also felt in forestry, fisheries, farming, and ranching. Combined cost estimates for all non-indigenous species in the United States are estimated at $137 billion per year (Pimentel et al. 2000).

Most exotic species do not become pests. In fact, of the 158 exotic plant species identified in our study areas, only 15 are considered pests of concern (Appendix A; Hogan and Huisinga 2000; B. Jacobs, personal communication, 18 Feb 2004; Romme et al. 2003a; Thomas et al. 2003; P. Whitefield, personal communication 2004). Some of these exotic species were introduced intentionally, as is the case with many European forage grasses introduced around the turn of the century. Others were introduced in shipments of wheat seed from Europe (probably the method by which cheatgrass, *Bromus tectorum*, was introduced), as weed seeds included in straw used during revegetation efforts, by livestock, on the wheels of vehicles, or in the shoe treds of hikers. Some exotic species have become sources of foods, herbs, craft items, and medicines to local people, and are sometimes incorporated into ceremonies. A recent wild foraging project in Flagstaff, Arizona, has targeted several wild exotic resources for gathering (P. West, personal communication, 3 March 2004). For example, purslane (*Portulaca oleracea*) is an invasive native plant in disturbed areas, but it is also a highly nutritious food that has medicinal uses (Moerman 1998). Similarly, Himalayan blackberry (*Rubus discolor*) is an invasive non-native plant, but the leaves are used as a tea and the fruits are also harvested (Moerman 1998).

Just as not all exotic species are invasive, not all invasive species are exotic. In many places, disturbed environments have proven favorable to a subset of native species that have expanded their populations to the detriment of other native species (Garrott et al. 1993). Sixty of the invasive species identified in our study areas are considered native. Appendix A lists invasive exotic and native plant species in our study areas.

Though invasive plant species can and do invade undisturbed areas of wilderness, invasion is more common in disturbed areas; indeed, disturbed habitats are increasingly vulnerable to invasion by exotic weeds (Floyd et al. 2001). With increasing fire frequencies (Grissino-Mayer and Swetnam 1997), more and more area in the West is being disturbed and opened to invasion simply by wildfire. In a 5-year post-fire study at Mesa Verde National Park, Adams (2002) noted that burned areas were very quickly invaded by exotic Russian thistle and prickly lettuce (*Lactuca serriola*). This increased threat of invasion is also noted in a recent plan for integrated treatment of noxious and invasive weeds in several counties and national forest areas in northern Arizona. Researchers note "dramatic increases over the past 30 years" in the land area covered by noxious or invasive plant species (Northern and Southern Colorado Plateau Networks 2004; Thomas et al. 2003). The worst problems are with bull thistle (*Cirsium vulgare*), leafy spurge (*Euphorbia escua*), various knapweed species, Dalmatian toadflax (*Linaria dalmatica*), and the woody species salt cedar (*Tamarix* sp.), Russian olive, and tree of heaven (*Ailanthus altissima*; Northern and Southern Colorado Plateau Networks 2004).

It is not surprising that invasive species have been identified as an important threat to biodiversity within three of our study areas. In fact, "invasive plant management is a priority natural resource problem for parks comprising the Northern and Southern Colorado Plateau monitoring networks" (Northern and Southern Colorado Plateau Networks 2004). Invasive species have been identified as a threat to spring-supported communities, hanging gardens, riparian zones, desert grasslands, and unusual or relict communities in Colorado Plateau national parks (Northern and Southern Colorado Plateau Networks 2004). The national parks and monuments on the Colorado Plateau are working on a unified management plan for combating invasive species, beginning with inventory and monitoring.

Invasive species constitute approximately 17 percent of the vascular plant species in Bandelier National Monument, but are not considered a serious management issue at present, partly because of relatively intact riparian zones on tributary creeks acting as a buffer (B. Jacobs, personal communication 2004). Exotics currently targeted for control at Bandelier include four woody species—salt cedar, Siberian elm (*Ulmus sibericia*), tree of heaven, and Russian olive—which can all be controlled with a mix of mechanical and chemical treatments (cut and then apply herbicide to the stump). Management of other exotics, which are fairly widespread throughout the monument, will be included in landscape-level habitat restoration projects designed to enhance opportunities for native species, with possible introduction of biological controls after further study (Jacobs 2002).

At Mesa Verde National Park, invasives have become a problem along riparian areas as well as in areas disturbed by wildfire. Eighty-seven non-native plant species—almost 14 percent of the total number of species in the park—have been identified. Romme et al. (2003a) have documented a marked increase in the degree to which exotics invade burned areas. Prior to the large fires of 1989, large wildfires were not associated with exotic weed invasions. After 1989, however, exotic invasive weeds became a serious problem following wildfires, which have disturbed approximately 28,340 acres in the park just during the past 7 years (www.nps.gov/meve/fire/). Though fires occurred in both pinyon-juniper and petran chaparral habitats, the former recovered much more quickly with native vegetation, and invasion by exotic species was not as much a problem as it was in the pinyon-juniper habitats. Efforts to stem the spread of invasives after fire have therefore been concentrated in the pinyon-juniper habitat (Romme et al. 2003a). The BAER recovery plan (burned areas emergency rehabilitation; National Park Service 2000) calls for reseeding burned areas with native grasses to help prevent establishment of invasive species after fires, a tactic that has proven effective. Experimental efforts from 1996 to 1998 to determine the effectiveness of various measures to combat invasive species have included aerial distribution of native grass seed over burned areas, mechanical removal of invading plants (specifically musk thistle), chemical removal (specifically for Canada thistle, *Cirsium arvense*), and release of biocontrol agents (specific to Canada thistle and musk thistle). These treatments met with some success, but did not prevent invasion and did not eliminate invaders (Floyd et al. 2001; Romme et al. 2003a).

Of the 838 species identified so far in Canyon de Chelly National Monument, 85, or about 10 percent, are exotic (Rink 2003). Of the 41 invasive plant species identified in the monument during a recent floristic study (Rink 2003), 30 were exotic but 11 were native invasive species. Invasive species of concern in Canyon de Chelly National Monument include salt cedar and Russian olive, which are spreading into Canyon del Muerto and Canyon de Chelly along riparian corridors. Russian olive was introduced to control erosion in the early 1900s (Thomas et al. 2003), and salt cedar was introduced in 1964 (Rink 2003); both were planted purposefully for erosion control (Rink 2003). Current management objectives for invasive species are focusing on a watershed approach that incorporates enhancement of sustainable agriculture as well as enhancing native biodiversity. Current plans call for experimental removal of salt cedar and Russian olive in some riparian areas (R. Hiebert, personal communication, 8 March 2004).

Wupatki National Monument protects one of the last remaining native grasslands in the Southwest that is not being grazed (though clearly it is not "pristine" because it was grazed until 1989); threats to this grassland include invasion by exotic species. Species of greatest concern include salt cedar, camelthorn (*Alhagi maurorum*), and cheatgrass. Riparian areas of the Little Colorado River corridor, an eastern boundary to the monument, are heavily invaded with salt cedar, and patches of camelthorn larger than 123 acres have been mapped in the monument using aerial photography interpretation and ground-truthing methods (Hansen 2002). Individual plants and small patches of salt cedar and camelthorn have also been located and mapped with a GPS unit along reaches of the Wupatki Basin drainages in the monument. Although control of these species is still in the planning stages (National Park Service 2002a, 2002b, 2003), it is likely that a combination of mechanical and chemical control will be used for salt cedar (cut and apply herbicide to stump), and that chemical control may be effective against camelthorn. Cheatgrass occurs mainly in disturbed areas along roadways and trails, but no effective control measure has been identified. Based on casual observations, cheatgrass has not invaded grassland-savannah areas that burned in 1995 and 2000 (P. Whitefield, personal communication 2004). Horehound (*Marrubium vulgare*), tumbleweed (*Salsola kali*), Mexican fireweed (*Kochia scoparoides*), and salt lover (*Halogeton glomeratus*) are also found in localized areas within the monument, and manual removal methods have been used to control some populations (Hogan and Huisinga 2000; P. Whitefield, personal communication 2004).

In summary, the presence of invasive exotic plants shifts native and domestic herbivore use, increases fire frequencies, changes erosion and deposition rates, and increases competition. In many cases, it also homogenizes the landscape, decreasing biodiversity. Management plans for forests and woodlands can no longer consider them inconsequential, especially because mechanized removal of thinned trees and controlled burns often increase the abundance of invasives.

CONCLUSIONS AND RECOMMENDATIONS

The last century of forest and woodland management on the Colorado Plateau has resulted in habitats that are increasingly homogeneous, more prone to high-severity stand-replacing fires, and less diverse in understory non-timber forest products. These trends have generated serious ecological, economic, and cultural consequences not only in the region, but across the United States. Causes of these trends include interactions among land-use history, climatic factors, physical characteristics of the landscapes, and the ecological relationships of predators, prey, protected livestock, and forage species.

Using four case studies as examples, we have demonstrated that each site has unique characteristics that must be taken into account when formulating management or restoration plans, and in most cases, each landscape has unique needs. We did find a few general patterns, including the decline in species richness of native plants and birds, with across-the-board increases in exotic invasive plants over the last century and a half at all sites. In most places we found reductions in fire frequencies induced by historic overgrazing (though the timing and severity differed considerably) that predated governmental fire suppression mandates. We documented a reduction in understory species richness and a trend toward more homogeneity in the biota, and a reduction in avian diversity and abundance, particularly in fire-suppressed doghair thickets of ponderosa pines and in savannas and pinyon-juniper woodlands where livestock were poorly managed. We found an increased probability that fires will cause property damage in the wildlands/urban interface; habitat fragmentation has already

aggravated other conservation problems in these areas including greatly decreased biodiversity. Finally, and perhaps most important, we note a reduction in the heterogeneity of cultural management strategies for forests and woodlands. U.S. Forest Service and National Park Service policies have spread through the region and were adopted by BIA resource managers, replacing Native American practices to varying degrees. These federal policies have had the effect of regularizing approaches to land management with one-size-fits-all policies. Reduction of Native American and Hispanic traditional use has helped create a homogenized landscape, with a loss of practices that formerly maintained a mosaic of habitats through the use of fire, small-scale farming, and gathering. Prohibiting their access to natural resources has led to a loss of traditional ecological knowledge among these cultural groups. Another important effect of this trend has been to focus management on timber or grazing resources, to the detriment of non-timber forest products, which are important contributors to biodiversity and ecosystem health. The stewardship promoted and maintained by these traditional uses is lost when traditional uses are denied.

Some of these trends were of course taken into account when formulating the Healthy Forests, Healthy Communities legislation that is now guiding forest and woodland management in the intermountain West. However, the interpretations of this report differ in several respects from the premises of that legislative package, at least as it has begun to be implemented by the U.S. Forest Service and some other agencies:

First, we acknowledge the tremendous range of forest and woodland habitats, and reject a one-size-fits-all management formula based on the concept of presettlement reference point conditions. We warn against applying the same management principles being refined for certain southwestern ponderosa pine forests (especially at their urban interfaces) to all forest and woodland types. In particular, these fire management protocols are a poor fit for most pinyon-juniper woodlands, in which stand-replacing fires have occurred at dramatically different frequencies across habitat types, temporally and spatially, and in which prescribed burns are likely to cause loss of the pinyon and juniper. In fact, they do not even serve all ponderosa forest types well.

Perhaps more pragmatically, we question whether industrial logging or medium-diameter thinning will result in true long-term restoration of forest health per se, as it addresses only forest structure and not the restoration of ecological processes key to forest and woodland health. There may be some congruence between the health of individual trees, reduced crown fire risk, and forest structure thinning, at least in some ponderosa forests (P. Friederici, personal communication 2004). However, recently thinned forests have not necessarily triggered recovery of biodiversity, or more particularly, understory species richness, within the first decade after treatment.

Many scientists have fundamental doubts that dendrochronological reconstructions of forest structure are sufficient to guide restoration and management, given that climate change, habitat fragmentation, pollution, loss of predators, and invasive exotic species have altered forest regeneration dynamics, and that these reconstructions tell us little about forest composition and ecological processes. In addition, dendrochronology, if used as the only tool for defining reference conditions, under-represents understory diversity and dynamics. In short, mono-causal explanations for habitat change (e.g., inappropriate grazing management) are not particularly useful. Our case studies demonstrate that land use interacts with other factors—especially climatic conditions and other localized conditions—to influence the direction and intensity of change.

Instead, we encourage more site-specific analyses of forest and woodland history, using packrat middens, historic photos, written records, pollen records, and phytolith analyses to obtain a richer perspective on past levels of biodiversity than dendrochronology alone will ever allow. Oral history and traditional ecological knowledge can also play vital roles in helping set restoration goals, but these goals cannot be simple reconstructions of the past. Careful assessments of climate change and recently arrived zoobiotics as well as invasive plants must be factored in.

Ultimately, we agree with those who promote a stronger focus on restoring ecological processes rather than exclusively focusing on structure or composition. All three are essential for healthy forest and woodland ecosystems. Managing for ecological integrity, heterogeneity, sustainability, and resilience are more important than attempting to return to some snapshot of presettlement conditions. That is why resource managers need decision-support research summaries that also consider pollution, climate change, and habitat fragmentation and loss.

Last, but not least, our discussions with forest practitioners from various cultures make us certain that the historic and contemporary uses of non-timber forest products have been undervalued relative to timber and livestock production goals by most managers on the Colorado Plateau. We applaud recent efforts to remove barriers to collaborative multicultural forestry (Moote and Becker 2003), but we believe that all forest management plans in our region need to more fully accommodate traditional as well as emerging uses of forests and woodlands in addition to timber, fuelwood, and livestock production. Otherwise, we will restore for conifers and grass, but not for harvestable understory species and for wildlife.

Given these concerns, we make the following recommendations to ensure that forest and woodland biodiversity can be restored. First, and most important, adaptively managing for uncertainty (drought and climate change) is preferable to managing for a single paradigm of timber, livestock, or fire management. Directing research toward restoring forest and woodland ecological processes, given current and predicted climatic regimes, should take priority over reconstructing past vegetation structure or composition. Current management is based on forestry or silviculture, with a focus on structure; perhaps a focus on ecosystem policy, emphasizing process, composition, and structure, would be more appropriate. It follows that livestock grazing and timber (or fuelwood) production optimization should not be the exclusive or even the primary goals; diverse vegetation patches in a heterogeneous mosaic will offer greater ecological integrity, resilience, and sustainability in the face of climatic and scientific uncertainty. Nor should wildfire reduction in the urban/wildlands interface take precedence over other human concerns. Preserving biodiversity and endemics, and creating and preserving culturally useful (and safe) landscapes, more likely than not will go hand in hand. The harvest of non-timber forest products used by diverse cultures is one way to help restore such a heterogeneous mosaic and increase bio-diversity. Harvesting of these products and goat or sheep foraging in the wildlands/urban interface can serve to reduce wildfire risk, while meeting other needs as well. These interests have not been adequately taken into account in past forest management plans; their inclusion will require stronger efforts toward implementing community-based forest planning that involves minority cultures and supports grassroots efforts to reinvigorate these traditions through language preservation and elder-youth mentoring programs.

Second, management plans for ponderosa pine forests and pinyon-juniper woodlands should be based as much as possible on site-specific information rather than on regional generalizations. Agencies must abandon more formulaic approaches and instead should base historic reconstructions of vegetation dynamics not merely on dendro-chronological samples and timber cutting records. As noted above, pollen, packrat middens, plant macrofossils, phytoliths, historic photos, archaeology, survey records, stocking rates, predator control data, and oral histories need to be integrated, where available. Managers should also consider the variety of current and predicted threats specific to their landscapes. Understandably, agencies may balk at what may seem an expensive and time-consuming assessment task. In these cases, collaborations between agencies and universities or other non-governmental organizations may provide possibilities for exchanges of information that will prove valuable to all stakeholders. Here at Northern Arizona University, we are fortunate to have one such collaborative arrangement, the Colorado Plateau Cooperative Ecosystem Studies Unit (CP/CESU) (http://cpcesu.nau.edu/new/index.htm), which brings agency staff and university researchers together to solve resource problems at multiple scales using interdisciplinary ecosystem studies. The CP/CESU brings representatives from five federal agencies—the USGS, BLM, Bureau of Reclamation, U.S. Forest Service, and NPS—together with 14 western universities and non-governmental organizations. The ability to access resources from such a diverse group has a positive effect for all parties involved. Even when collaborations of this type are not in place, consultation with re-searchers interested in paleoenvironmental studies can prove fruitful. Additionally, using forest records and oral histories with retired forest managers can be productive.

Third, given the recognition of invasive plant species as a significant threat over much of the Colorado Plateau—and indeed, across the United States—forest thinning methods that increase the dispersal, recruitment, and establishment of invasive exotic weeds should be strongly avoided. Because most landscapes on the Colorado Plateau have evolved with low levels of soil surface disturbance, and are dependent on the biological crusts that have developed, management and restoration efforts should minimize disturbance of soil crusts in thinning and burning activities, and in permitting recreational use, logging, mining, livestock management, and other activities that would disturb the soil's biological crusts.

Finally, as public awareness of forest management increases, ecological restoration must be presented to the public as a set of management strategies broader than structural thinning and controlled burns, or restorationists will win short-term battles while losing "the war." Managing our forest and woodland ecosystems for commodities such as lumber and livestock production will result in a narrow set of stakeholders, who are often not part of the local economy. Shifting our focus to managing them as ecosystems, with cultural and economic ties to the local population, will create a broader and more engaged group of stakeholders who have the health of the ecosystem in mind. We realize that it may be difficult to promote these practices, which are intended to restore the overall health to our western ecosystems, during this time of heightened fear of wildfire. Increasing a sense of stewardship through education and a stronger focus on other uses of the forest, such as harvest of non-timber forest products, traditional uses, hunting, and wildlife observation and photography, will be invaluable tools for managers willing to take these important first steps.

LITERATURE CITED

Abruzzi, W. S. 1985. Water and community development in the Little Colorado River basin. Human Ecology 12: 241–269.

Abruzzi, W. S. 1989. Ecology, resource redistribution and Mormon settlement in northeastern Arizona. American Anthropologist 91: 642–655.

Abruzzi, W. S. 1995. The social and ecological consequences of early cattle ranching in northeastern Arizona. Human Ecology 23: 75–98.

Adams, K. R. 2002. A five-year ecological and ethnobotanical study of vegetation recovery after the Long Mesa Fire of July 1989, Mesa Verde National Park, Colorado. Manuscript on file, Mesa Verde National Park.

Adams, K.R. 2002. A Five-year ecological and ethnobotanical study of vegetation recovery after the Long Mesa Fire of July 1989, Mesa Verde National Park, Colorado. Manuscript on file, Mesa Verde National Park.

Adams, K. R., and K. L. Petersen. 1999. Environment. In Colorado prehistory: A context for the southern Colorado River basin, edited by W. D. Lipe, M. D. Varien, and R. H. Wilshusen, pp. 14–50. Colorado Council of Professional Archaeologists, Denver.

Ahlstrom, R. V. N., C. R. Van West, and J. S. Dean. 1995. Environmental and chronological factors in the Mesa Verde–Northern Rio Grande migration. Journal of Anthropological Archaeology 14: 125–142.

Alcoze, T. 2003. First peoples of the pines: Historical ecology of humans and ponderosas. In Ecological restoration of southwestern ponderosa pine forests, edited by P. Friederici, pp. 48–57. Island Press, Washington DC.

Alcoze, T., and M. Hurteau. 2001. Implementing the archaeo-environmental reconstruction technique: Rediscovering the historic ground layer of three plant communities in the Great Grand Canyon region. In Historical ecological handbook, edited by D. Egan and E. A. Howell, pp. 413–425. Island Press, Washington DC.

Allen, C. D. 1989. Changes in the landscape of the Jemez Mountains, New Mexico. Unpublished Ph.D. dissertation, University of California, Berkeley.

Allen, C. D. 1998. Where have all the grasslands gone? Quivera Coalition Newsletter, Spring/Summer.

Allen, C. D. 2002a. Lots of lightning and plenty of people. An ecological history of fire in the upland Southwest. In Fire, native peoples, and the natural landscape, edited by T. R. Vale, pp. 143–193. Island Press, Washington DC.

Allen, C. D. 2002b. Rumblings in the Rio Arriba: Landscape changes in the southern Rocky Mountains of northern New Mexico. In Rocky Mountain futures. An ecological perspective, edited by J. S. Baron, pp. 239–253. Island Press, Washington DC.

Allen, C. D. 2004. Ecological patterns and environmental change in the Bandelier landscape. In Archaeology of Bandelier National Monument: Village formation on the Pajarito Plateau, New Mexico, edited by T. A. Kohler. University of New Mexico Press, Albuquerque.

Allen, C. D., and D. D. Breshears. 1998. Drought-induced shift of a forest/woodland ecotone: Rapid landscape response to climate variation. Proceedings of the National Academy of Sciences of the United States of America 95: 14839–14842.

Allen, C. D., J. L. Betancourt, and T. W. Swetnam. 1998. Landscape changes in the southwestern United States: Techniques, long-term data sets, and trends. In Perspectives on the land use history of North America: A context for understanding our changing environment. Biological Science Report USGS/BRD/BSR-1998-0003.

Allen, C. D., D. A. Falk, M. Hoffman, J. Klingel, P. Morgan, M. Savage, T. Schulke, P. Stacey, K. Suckling, and T. W. Swetman. 2002. Ecological restoration of southwestern ponderosa pine ecosystems: A broad framework. Ecological Applications 12: 1418–1433.

Allen, C. D., K. L. Beeley, and T. W. Swetnam. 2003. Southwestern drought and landscape-scale vegetation dieback: The 1950s and now. Seventh biennial conference of research on the Colorado Plateau, programs and abstracts, p. 23.

Altman, D. G., I. Chalmers, M. Egger, G. Smith, and G. Davey. 2001. Systematic reviews in health care: Meta-analysis in context. BMJ Books, London. Ellis, F. H. 1974. Extensions of land use: Inner to outer areas—Animals and birds utilized. In The Hopi: Their history and use, edited by D. A. Horr, pp. 145–187. Hopi Indians in American Indian Ethnohistory. Garland, New York.

Anderson, B. A. 1990. Chapter 3. In The Wupatki Archaeological Inventory Survey Project: Final report, edited by B. A. Anderson, pp. 1–38. Southwest Cultural Resources Professional Paper 35. Santa Fe NM.

Anderson, B. A. 1993. Wupatki National Monument: Exploring into prehistory. In Wupatki and Walnut Canyon: New perspectives on history, prehistory and rock art, edited by D. G. Noble, pp. 13–19. Ancient City Press, Santa Fe, NM.

Anderson, M. K. 1991. California Indian horticulture. Fremontia 18(2): 7–14.

Anderson, R. S. 1989. Development of the southwestern ponderosa pine forests. What do we really know? Paper presented at Multiresource Management of Ponderosa Pine Forest Symposium. Northern Arizona University, Flagstaff, November 14–16, 1989.

Anderson, R. S. 1993. A 35,000-year vegetation and climate history from Potato Lake, Mogollon Rim, Arizona. Quaternary Research 40: 351–359.

Anderson, R. S., C. D. Allen, J. Toney, R. Jass, and A. Bair. 2003a. Holocene vegetation and forest fire regimes in subalpine and mixed conifer forests, southern Colorado and northern New Mexico. Seventh Biennial Conference of Research on the Colorado Plateau, programs and abstracts, p. 23.

Anderson, J., J. Anhold, J. McMillin, and N. Cobb. 2003b. Regional and local patterns of bark beetle induced tree mortality on the Colorado Plateau. Seventh Biennial Conference of Research on the Colorado Plateau, programs and abstracts, p. 23.

Anhold, J. A., R. Fitzgibbon, M. L. Fairweather, J. McMillin, S. Dudley, D. Allen-Reid, T. Rogers, D. Conklin, and R. Norris. 2001. Insect and disease conditions in the southwestern region, 2000. Southwest Region Forestry and Forest Health. U.S. Department of Agriculture, Forest Service, Southwestern Region, R3-00-01.

Bahre, C. 1991. A legacy of change: Historic human impacts on vegetation in the Arizona borderlands. Unviversity of Arizona Press, Tucson.

Bailey, G., and R. G. Bailey. 1986. A history of the Navajos: The reservation years. School of American Research Press, Santa Fe NM.

Bailey, R. G., P. E. Avers, T. King, and W. H. McNab, Editors. 1994. Ecoregions and subregions of the United States (map). Washington, D.C.: USDA Forest Service. 1: 7,5000,000. With supplementary table of map unit descriptions, compiled and edited by W.H. McNab and R.G. Bailey.

Baker, B. 2001. National management plan for controlling invasive species. BioScience 51: 92.

Baker, R. D., R. S. Maxwell, V. H. Treat, and H. C. Dethloff. 1988. Timeless heritage: A history of the Forest Service in the Southwest. U.S. Department of Agriculture Forest Service FS-409, Washington DC.

Baker, W. L., and D. J. Shinneman. 2004. Fire and restoration of piñon-juniper woodlands in the western United States: A review. Forest Ecology and Management 189(1–3): 1–21.

Baker, W. L., and T. T. Veblen. 1990. Spruce beetles and fires in the 19th century subalpine forests of western Colorado, USA. Arctic and Alpine Research 22(1): 65–80.

Balice, R. G., S. G. Ferran, and T. S. Foxx. 1997. Preliminary vegetation and land cover classification for the Los Alamos region. Los Alamos National Laboratory LA-UR-97-4627. Los Alamos NM.

Barnola, J.-M., D. Raynaud, C. Lorius, and N. I. Barkov. 2003. Historical CO_2 record from the Vostok ice core. In Trends: A compendium of data on global change. Carbon Dioxide Information Analysis Center, Oak Ridge National Laboratory, U.S. Department of Energy, Oak Ridge TN. http://cdiac.esd.ornl.gov/trends/co2/vostok.htm

Bartolome, J. W. 1993. Application of herbivore optimization theory to rangelands of the western United States. Ecological Applications 3: 27–29.

Baxter, J. O. 1987. Las Carneradas: Sheep trade in New Mexico, 1700–1860. University of New Mexico Press, Albuquerque.

Baydo, G. R. 1970. Cattle ranching in territorial New Mexico. Ph.D. dissertation, University of New Mexico, Albuquerque.

Belnap, J. 1995. Surface disturbances: Their role in accelerating desertification. Environmental Monitoring and Assessment 37: 39–57.

Belnap, J. 2003. Magnificent microbes: Biological soil crusts in pinyon-juniper communities. In Ancient pinyon-juniper woodlands: A natural history of Mesa Verde country, edited by M. L. Floyd, pp. 75–88. University Press of Colorado, Boulder.

Belsky, A. J., A. Matzke, and S. Uselman. 1999. Survey of livestock influences on stream and riparian ecosystems in the western United States. Journal of Soul and Water Conservation 54(1): 419–431.

Bennett, D., B. Tkacz, J. Wilson, M. Fairweather, T. Rogers, D. Conklin, and M. Schultz. 1994. Insects and disease: Indicators of forest health, a forest health briefing paper. Forest Pest Management Report R3-95-02. USDA Forest Service, Southwestern Region, Albuquerque NM.

Berlin, G. Lennis, D. E. Salas, and P. R. Geib. 1990. A prehistoric Sinagua agricultural site in the ashfall zone of Sunset Crater, Arizona. Journal of Field Archaeology 17(1): 1–16.

Berry, M. S. 1982. Time, space, and transition in Anasazi prehistory. University of Utah Press, Salt Lake City.

Betancourt, J. L. 1990. Late Quaternary biogeography of the Colorado Plateau. In Packrat middens. The last 40,000 years of biotic change, edited by J. L. Betancourt, T. R. Van Devender, and P. S. Martin, pp. 259–292. University of Arizona Press, Tucson.

Betancourt, J. L., and O. K. Davis. 1984. Packrat middens from Canyon de Chelly, northeastern Arizona: Paleoecological and archaeological implications. Quaternary Research 21: 56–64.

Betancourt, J. L., and T. R. Van Devender. 1981. Holocene vegetation in Chaco Canyon, New Mexico. Science 214: 656–658.

Betancourt, J. L., and T. R. Van Devender. 1983. Fossil packrat middens from Chaco Canyon, New Mexico: Cultural and ecological significance. In Chaco Canyon country: A field guide to the geomorphology, Quaternary geology, paleoecology, and environmental geology of northwestern New Mexico, edited by S. G. Well, D. Love, and T. W. Gardner, pp. 207–217. 1983 Field Trip Guidebook. American Geomorphology Field Group.

Betancourt, J. L., J. S. Dean, and H. M. Hull. 1986. Prehistoric long-distance transport of construction beams, Chaco Canyon, New Mexico. American Antiquity 51(2): 370–375.

Betancourt, J. L., T. R. Van Devender, and P. S. Martin. 1990. Synthesis and prospectus. In Packrat middens. The last 40,000 years of biotic change, edited by J. L. Betancourt, T. R. Van Devender, and P. S. Martin, pp. 435–447. University of Arizona Press, Tucson.

Binkley, D., M. M. Moore, W. H. Romme, and P. M. Brown. In review. Was Aldo Leopold right about the Kaibab deer herd? Submitted to Ecosystems.

Birdsey, R., R. Alig, and D. Adams. 2000. Mitigation activities in the forest sector to reduce emissions and enhance sinks of greenhouse gases. In The impact of climate change on America's forests: A technical document supporting the 2000 USDA Forest Service RPA assessment, technical editors L. A. Joyce and R. Birdsey. USDA Forest Service, Rocky Mountain Research Station, General Technical Report RMRS-GTR-59. Fort Collins CO.

Blinn, D. W., R. H. Hevly, and O. K. Davis. 1994. Continuous Holocene record of diatom stratigraphy, paleohydrology, and anthropogenic activity in a spring-mound in southwestern United States. Quaternary Research 42: 197–205.

Blythe, C. P. 1995. The Cenozoic evolution of Wupatki National Monument. Master's thesis on file at Northern Arizona University, Flagstaff.

Bohrer, V. L. 1975. The prehistoric and historic role of the cool-season grasses in the Southwest. Economic Botany 29: 199–207.

Bohrer, V. L. 1983. New life from ashes: The tale of the burnt bush (*Rhus trilobata*). Desert Plants 5(3): 122–124.

Bostick, V. B. 1949. North Kaibab deer-livestock forage relationship study. Progress report (September 15, 1949). USDA Forest Service, Kaibab National Forest, Arizona.

Botkin, D. 1990. Doscordant harmonies: A new ecology for the twenty-first century. Oxford University Press, New York.

Bowen, B. M. 1990. Los Alamos climatology. Los Alamos National Laboratory LA-11735-MS UC-902, Los Alamos NM.

Brandt, C. 1998. Analysis of plant macro-remains. In The N30-31 project: Investigations at twenty-two sites between Mexican Springs and Navajo, McKinley County, Vol. 3, edited by J. E. Damp and E. J. Skinner, pp. 133–184. Research Series 10. Zuni Cultural Resource Enterprise, Zuni NM.

Brandt, C. B. 1999. Analysis of plant microremains. In Chuska chronologies, houses, and hogans: Archaeological and ethnographic inquiry along N30-N31 between Mexican Springs and Navajo, McKinley County, New Mexico, Volume III, Part 2: Analysis, edited by J. E. Damp, pp. 441–492. Zuni Archaeological Program, Pueblo of Zuni.

Brawn, J. D., and R. P. Balda. 1988. The influence of silviculture activity on ponderosa pine forest bird communities in the southwestern United States. In Bird conservation, edited by J. A. Jackson, pp. 3–21. University of Wisconsin Press, Madison.

Breshears, D. D., and C. D. Allen. 2002. The importance of rapid, disturbance-induced losses in carbon management and sequestration. Global Ecology and Biogeography 11: 1–5.

Brown, D. E. 1994. Biotic communities: Southwestern United States and northwestern Mexico. University of Utah Press, Salt Lake City.

Brown, G. M. 1996. The protohistoric transition in the northern San Juan region. In The archaeology of Navajo origins, edited by R. H. Towner, pp. 47–69. University of Utah Press, Salt Lake City.

Brown, J. H., T. G. Whitham, S. K. Morgan Ernest, and C. A. Gehring. 2001. Complex species interactions and the dynamics of ecological systems: Long-term experiments. Science 293 (5530): 643–649.

Brugge, D. M., and R. Wilson. 1976 (updated March 2004). Administrative history: Canyon de Chelly National Monument, Arizona. U.S. Department of Interior, National Park Service. http://www.nps.gov/cach/adhi/adhi.htm.

Brunner-Jass, R. 1999. Fire occurrence and paleoecology at Alamo Bog and Chihuahueños Bog, Jemez Mountains, New Mexico, USA. Master's thesis, Northern Arizona University, Flagstaff.

Bunting, S. C. 1986. Use of prescribed burning in juniper and pinyon-juniper woodlands. In Proceedings – Pinyon-Juniper Conference, Reno NV, January 13–16, 1986, edited by R. L. Everett, pp. 141–144. Intermountain Research Station, Ogden UT.

Burgess, T. L. 1973. Mammals of the Canyon de Chelly region, Apache Co., Arizona. Journal of Arizona Academy of Science 8: 21–25.

Burk, J. C. 1973. The Kaibab deer incident: A long persisting myth. BioScience 9: 43–47.

Burnham, P. 2000. Indian country – God's country: Native Americans and the national parks. Island Press, Washington DC.

Burns, B. T. 1983. Simulated Anasazi storage behavior using crop yields reconstructed from tree rings: A.D. 652–1968. Ph.D. dissertation, University of Arizona. University Microfilms International, Ann Arbor.

Carlson, A. W. 1969. New Mexico's sheep industry, 1850–1900: Its role in the history of the territory. New Mexico Historical Review XLIV 1: 25–49.

Caughley, G. 1970. Eruption of ungulate populations with emphasis on Himalayan thar in New Zealand. Ecology 51: 53–71.

Cayan, D. R., K. T. Redmond, and L. G. Riddle. 1999. ENSO and hydrologic extremes in the western United States. Journal of Climate (12): 2881–2893.

Christensen, N. L. 1989. Landscape history and ecological change. Journal of Forest History 33:116–124.

Christman, C. J., D. P. Sponeberg, and D. E. Bixby. 1997. Rare breeds album of American livestock. American Livestock Breeds Conservancy, Pittsboro NC.

Cinnamon, S. K. 1981. Livestock grazing at Wupatki National Monument. Unpublished manuscript on file at Wupatki National Monument. 16 pp.

Cinnamon, S. K. 1988. The vegetation community of Cedar Canyon, Wupatki National Monument as influenced by prehistoric and historic environmental change. Master's thesis, Northern Arizona University, Flagstaff.

Cobb, N. S., S. Mopper, C. A. Gehring, M. Caouette, K. M. Christensen, and T. G. Whitham. 1997. Increased moth herbivory associated with environmental stress of pinyon pine at local and regional levels. Oecologia 109: 389–397.

Coder, C. M., V. Randall, E. Smith-Rocha, and R. Hines. 2004. *Chi ch'il* (acorns): Dissolution of traditional Dilzhe'e gathering practices due to federal control of the landscape. Paper presented at Biodiversity and Management of the Madrean Archipelago II: Connecting Mountain Islands and Desert Seas, May 11–14, 2004. Tucson AZ.

COHMAP. 1988. Climatic changes of the last 18,000 years: Observations and model simulation. Science 241: 1043–1052.

Cole, K. L. 1990. Late Quaternary vegetation gradients through the Grand Canyon. In Packrat middens. The last 40,000 years of biotic change, edited by J. L. Betancourt, T. R. Van Devender, and P. S. Martin, pp. 240–258. University of Arizona Press, Tucson.

Cole, K. L., N. Henderson, and D. S. Shafer. 1997. Holocene vegetation and historic grazing impacts at Capitol Reef National Park reconstructed from packrat middens. Great Basin Naturalist 57: 315–326.

Colton, H. S. 1932. Sunset Crater: The effect of a volcanic eruption on an ancient Pueblo people. The Geographical Review 22(4): 582–950.

Colton, H. S. 1960. Black sand: Prehistory in Northern Arizona. University of New Mexico Press, Albuquerque.

Colyer, M. 2003. Some ethnobotanical uses of plants from the pinyon-juniper woodlands. In Ancient pinyon-juniper woodlands: A natural history of Mesa Verde country, edited by M. L. Floyd, pp. 295–308. University of Colorado Press, Boulder.

Conklin, D. A. 2004. Development of the white pine blister rust outbreak in New Mexico. USDA Forest Service R3-04-01. http://www.fs.fed.us/r3/publications/documents/wp_blister_rust_nm.pdf

Cordell, L. S. 1994. Ancient pueblo peoples. Smithsonian Institution, Washington DC.

Couzin, J. 1999. Landscape changes make regional climate run hot and cold. Science 283: 317–318.

Covington, W. W. 2003. The evolutionary and historical context. In Ecological restoration of southwestern ponderosa pine forests, edited by P. Friederici, pp. 26–47. Island Press, Washington DC.

Covington, W. W., and M. M. Moore. 1994. Southwestern ponderosa pine forest structure: Changes since Euro-American settlement. Journal of Forestry 92(1): 39–47.

Covington, W. W., P. Z. Fulé, M. M. Moore, S. C. Hart, T. E. Kolb, J. N. Mast, S. S. Sackett, and M. R. Wagner. 1997. Restoring ecosystem health in ponderosa pine forests of the Southwest. Journal of Forestry 95(4): 23–29.

Crooks, K. R., and M. E. Soulé. 1999. Mesopredator release and avifaunal extinctions in a fragmented system. Nature 400: 563–566.

Dahms, C. W., and B. W. Geils, Editors. 1997. An assessment of forest ecosystem health in the Southwest. USDA Forest Service Rocky Mountain Forest and Range Experiment Station Southwestern Region, General Technical Report RM-GTR-295.

Dansgaard, W., J. W. C. White, and S. J. Johnsen. 1989. The abrupt termination of the younger dryas climate event. Science 339: 532–534.

D'Antonio, C., and P. M. Vitousek. 1992. Biological invasions by exotic grasses, the grass/fire cycle, and global change. Annual Review of Ecology and Systematics 23: 63–87.

Davenport, D. W., D. D. Breshears, B. P Wilcox, and C. D. Allen. 1998. Viewpoint: Sustainability of pinyon-juniper ecosystems—A unifying perspective of soil erosion thresholds. Journal of Range Management 51: 231–240.

Davidson-Hunt, I., L. C. Duchesne, and J. C. Sasada. 2001. Non-timber forest products: Local livelihoods and integrated forest management. In Forest communities in the third millennium: Linking research, business, and policy toward a sustainable non-timber forest product sector. Proceedings of meeting held October 1–4, 1999, Ontario, Canada. USDA Forest Service, North Central Research Station, St. Paul. GTR-NC-217.

Dean, G. 1998. Pollen evidence of human activities in the southern Chuska Mountains from Basketmaker III through Historic Navajo. In The N30-31 project: Investigations at twenty-two sites between Mexican Springs and Navajo, McKinley County, Vol. 3, edited by J. E. Damp and E. J. Skinner, pp. 185–294. Research Series 10. Zuni Cultural Resource Enterprise, Zuni NM.

Dean, J. S. 1988. Dendrochronology and paleoenvironmental reconstruction on the Colorado plateaus. In The Anasazi in a changing environment, edited by G. J. Gumerman, pp. 119–167. Cambridge University Press, Cambridge UK.

Dean, J. S., and G. S. Funkhouser. 1995. Dendroclimatic reconstructions for the southern Colorado Plateau. In Proceeding of a workshop on climate change in the four corners and adjacent regions, edited by W. J. Waugh, pp. 85–104. U.S. Department of Energy, Grand Junction Projects Office, Grand Junction CO.

Dean, W. E., T. S. Ahlbrandt, R. Y. Anderson, and J. P. Bradbury. 1996. Regional aridity in North America during the middle Holocene. The Holocene 6(2): 145–155.

De Gomez, T. 2003. Pine bark beetle outbreak in Arizona. University of Arizona Press Release, Tucson.

De Gomez, T., and C. Young 2002. Pine bark beetles. University of Arizona Cooperative Extension Publication AZ1300, Tucson.

Deichmann, J. W. 1980. Botanical survey of the Monument Canyon Research Natural Area. USDA Forest Service Final Report on Contract 53-32-FT-8-22, Alburquerque.

Dennis, A. E. 1975. The natural vegetation of Canyon de Chelly National Monument. Kiva 41: 15–22.

Despain, D. W., and J. C. Mosley. 1990. Fire history and stand structure of a pinyon-juniper woodland at Walnut Canyon National Monument, Arizona. National Park Service Technical Report 34. Cooperative Park Studies Unit, Flagstaff AZ.

Diamond, J. 1986. The environmentalist myth: Archaeology. Nature 324: 19–30.

Diaz, H. F., and V. Markgraf, Editors. 2000. El Niño and the Southern Oscillation: Multiscale variability and global and regional impacts. Cambridge University Press, Cambridge UK.

Dieterich, J. J. 1980. Chimney Spring forest fire history. U.S. Department of Agriculture, Forest Service General Technical Report RM-220. Fort Collins CO.

Dobyns, H. 1981. Fire and flood. Ballena Press Anthropological Papers 20, Soccoro NM.

Dorman, R. L. 1996. The chili line and Santa Fe the city different. R.D. Publications, Santa Fe.

Downum, C. E. 1988. One grand history: A critical review of Flagstaff archaeology, 1851–1988. Ph.D. dissertation, University of Arizona, Tucson.

Downum, C. E., and A. P. Sullivan. 1990. Chapter 5. In The Wupatki Archaeological Inventory Survey Project: Final report, edited by B. Anderson, pp. 1–38. Southwest Cultural Resources Professional Paper 35. Santa Fe NM.

Dunmire, W. W., and G. D. Tierney. 1995. Wild plants of the pueblo province. Exploring ancient and enduring uses. Museum of New Mexico Press, Albuquerque.

Easterling, D. R., G. A. Meehl, C. Parmesan, S. A. Changnon, R. R. Karl, and L. O. Mearns. 2000. Climate extremes: Observations, modeling, and impacts. Science 289: 2068–2074.

Ecological Restoration Institute. 2002. Understory plant community restoration in the Unikaret Mountains, Arizona. Ecological Restoration Institute Working Papers in Southwest Ponderosa Forest Restoration 2, Northern Arizona University, Flagstaff.

Egan, D., and E. A. Howell. 2001. The historical ecology handbook. Island Press, Washington DC.

English, N. E., J. L. Betancourt, J. S. Dean, and J. Quade. 2001. Strontium isotopes reveal source of architectural timber at Chaco Canyon, New Mexico. Proceedings of the National Academy of Sciences 98, 11891–11896.

Erdman, J. A. 1970. Pinyon-juniper succession after natural fires on residual soils of Mesa Verde, Colorado. Brigham Young University Science Bulletin, Biological Series 11(2): 1–26.

Erdman, J. A., C. L. Douglas, and J. W. Marr. 1969. The environment of Mesa Verde, Colorado. Archaeological Research Series 7-B. National Park Service, Washington D.C.

Evans, R. D., and J. Belnap. 1999. Long-term consequences of disturbance on nitrogen dynamics in an arid ecosystem. Ecology 80(1): 150–160.

Evans, W. G. 1973. Fire beetles and forest fires. Insect World Digest 1: 14–18.

Everett, Y. 1997. A guide to selected non-timber forest products of the Hayfork Adaptive Management area, Shasta-Trinity and Six Rivers National Forests, California. USDA Forest Service General Technical Report PSW-GTR-162.

Fall, P. L., G. Kelson, and V. Markgraf. 1981. Paleoenvironmental reconstruction at Canyon del Muerto, Arizona, based on principal-component analysis. Journal of Archaeological Science 8: 2997–2307.

First Approximation Report (FAR). 2003. Oregon Department of Forestry. About United States 2003 Report on Sustainable Forests by W.B. Smith. August 2003. (www.fs.fed.us/research/sustain/pdfs/c6_i34.pdf).

Flannigan, M. D., Y. Bergeron, O. Engelmark, and B. M. Wotton. 1998. Future wildfire in circumboreal forests in relation to global warming. Journal of Vegetation Science 9: 469–476.

Fleischner, T. L. 1994. Ecological costs of livestock grazing in western North America. Conservation Biology 8(3): 629–644.

Floyd, L., Editor. 2003. Ancient pinyon-juniper woodlands: A natural history of Mesa Verde country. University of Colorado Press, Boulder.

Floyd, M. L., and M. Colyer. 2003. Beneath the trees: Shrubs, herbs, and some surprising rarities. In Ancient pinyon-juniper woodlands: A natural history of Mesa Verde country, edited by M. L. Floyd, pp. 31–60. University of Colorado Press, Boulder.

Floyd, M. L., D. D. Hanna, and G. Salamancha. 2001. Post-fire treatment of noxious weeds in Mesa Verde National Park, Colorado. Proceedings of the Fifth Biennial Conference on Research on the Colorado Plateau, edited by C. vanRiper III, K. A. Thomas, and M. A. Stuart. U.S. Geological Survey/FRESC Report Series USGSFRESC/COPL/2001/24.

Floyd, M. L., W. H. Romme, and D. D. Hanna. 2003a. Fire history and woodland structure in pinyon-juniper woodlands on Mesa Verde: Implications for management. Paper presented at Seventh Biennial Conference of Research on the Colorado Plateau, Northern Arizona University, Flagstaff.

Floyd, M. L., W. H. Romme, and D. D. Hanna. 2003b. Fire history. In Ancient pinyon-juniper woodlands: A natural history of Mesa Verde country, edited by M. L. Floyd, pp. 261–277. University of Colorado Press, Boulder.

Floyd, M. L., T. L. Fleischner, D. Hanna, and P. Whitefield. 2003c. Effects of historic livestock grazing on vegetation at Chaco Culture National Historic Park, New Mexico. Conservation Biology 17(6): 1703–1711.

Floyd, M. L., M. Colyer, D. D. Hanna, and W. H. Romme. 2003d. Gnarly old trees: Canopy characteristics of old-growth pinyon-juniper woodlands. In Ancient pinyon-juniper woodlands: A natural history of Mesa Verde country, edited by M. L. Floyd, pp. 11–30. University Press of Colorado, Boulder.

Floyd, M. L., D. D Hanna, and W. H. Romme. 2004. Historical and recent fire regimes in pinyon-juniper woodlands on Mesa Verde, Colorado, USA. Forest Ecology and Management 198: 269–289.

Floyd-Hanna, L., and W. H. Romme. 1995. Fire history and fire effects of Mesa Verde National Park: Final report. MEVE-R91-0160, May 12.

Ford, R. I. 1979. Paleoethnobotany in American archeology. In Advances in Archaeological Method and Theory, Vol. 2, pp. 285–336. Academic Press. Orlando FL.

Fowler, D. D., and C. S. Fowler, Editors. 1971. Anthropology of the Numa: John Wesley Powell's manuscripts on the Numic peoples of western North America 1868–1880. Smithsonian Contributions to Anthropology 14.

Foxx, T. S. 2002. Draft. Ecosystems of the Pajarito Plateau and East Jemez Mountains: Linking land and people. LANL Cultural Resources Management Plan. RRES-ECO, Los Alamos National Laboratory NM.

Foxx, T. S., and G. D. Tierney. 1985. Status of the flora of the Los Alamos National Environmental Research Park. Checklist of vascular plants of the Pajarito Plateau and Jemez Mountains. Los Alamos National Laboratory LA-8050-NERP, Vol. III. Los Alamos NM.

Foxx, T. S., G. D. Tierney, M. Mullen, and M. Salisbury. 1997. Old-field plant succession on the Pajarito Plateau. Los Alamos National Laboratory LA-13350. Los Alamos NM.

Friederici, P., Editor. 2003a. Ecological restoration of southwestern ponderosa pine forests. Island Press, Washington D.C.

Friederici, P. 2003b. The "Flagstaff model." In Ecological restoration of southwestern ponderosa pine forests, edited by P. Friederici, pp. 7–25. Island Press, Washington DC.

Fulé, P. Z., G. Verkamp, A. E. M. Waltz, and W. W. Covington. 2002. Burning under old-growth ponderosa pines on lava soils. Ecological Restoration Institute 62(3): 47–49.

Ganz, D. J., D. L. Dahlsten, and P. J. Shea. 2003. The post-burning response of bark beetles to prescribed burning treatments. USDA Forest Service Proceedings RMRS-P-29.

Garrott, R. G., P. J. White, and C. A. Vanderbilt White. 1993. Overabundance: An issue for conservation biologists? Conservation Biology 7: 946–949.

Gehring, C. A., and T. G. Whitham. 1994. Comparisons of ectomycorrhizae on pinyon pine (Pinus edulis; Pinaceae) across extremes of soil type and herbivory. American Journal of Botany 81: 1509–1516.

Gehring, C. A., T. Theimer, T. G. Whitham, and P. Keim. 1998. Ectomycorrhizal fungal community structure of pinyon pines growing in two environmental extremes. Ecology 79(5): 1562–1572.

Goff, F., B. S. Kues, M. A. Rogers, L. D. McFadden, and J. N. Gardner, Editors. 1996. The Jemez Mountains region. New Mexico Geological Society Guidebook, 47th Field Conference. Albuquerque NM.

Graybill, D. A., and M. Rose. 1985. Water: Managing a precious resource. Arizona Land and People 36: 3–4.

Griffitts, M. 2003. Bedrock geology. In Ancient pinyon-juniper woodlands: A natural history of Mesa Verde country, edited by M. L. Floyd, pp. 183–196. University of Colorado Press, Boulder.

Griggs, A. B., Managing Editor. 2001. Cerro Grande canyons of fire, spirit of community. Los Alamos National Laboratory, published by Los Alamos National Bank, NM.

Griggs, R. L. 1964. Geology and ground-water resources of the Los Alamos area, New Mexico. U.S. Geological Survey, Water Supply Paper 1753. Washington DC.

Grissino-Mayer, H. D. 1995. Tree-ring reconstruction of climate and fire history at El Malpais National Monument, New Mexico. Ph.D. dissertation, University of Arizona, Tucson.

Grissino-Mayer, H. D., and T. W. Swetnam. 1997. Multi-century history of wildfire in the ponderosa pine forests of El Malpais National Monument. New Mexico Bureau of Mines and Mineral Resources Bulletin 156: 163–171.

Grissino-Mayer, H. D., and T. W. Swetnam. 2000. Century-scale climate forcing of fire regimes in the American Southwest. The Holocene 10(2): 213–220.

Hammer, C. U., H. B. Clausen, and W. Dansgaard. 1980. Greenland ice sheet evidence of post-glacial volcanism and its climatic impact. Nature 288: 230–235.

Hansen, M. L. 2002. Biological and cultural sustainability of *Petrophytum caespitosum* at Walnut Canyon National Monument, Arizona. Masters thesis, Northern Arizona University, Flagstaff.

Harlan, A. S., and A. E. Dennis. 1976. A preliminary plant geography of Canyon de Chelly National Monument. Journal of the Arizona-Nevada Academy of Science 11(171): 69–78.

Hartman, R. L. 2003. Preliminary checklist of the vascular plants of Valles Caldera National Preserve, New Mexico. Manuscript on file, Bandelier National Monument, NM.

Healthy Forests. 2002. An initiative for wildfire prevention and stronger communities. Office of the President of the United States, August 22, 2002.

Henke, S. E., and F. C. Bryant. 1999. Effects of coyote removal on the faunal community in western Texas. Journal of Wildlife Management 63(4): 1066–1081.

Hereford, R., R. H. Webb, and S. Graham. 2000. Precipitation history of the Colorado Plateau region, 1900–2000. USGS Fact Sheet 119-02.

Hogan, P., and K. Huisinga. 2000. The Wupatki herbarium: Clyde Peshlakai ethnobotanical collection and Sunset Crater herbarium. Manuscript on file, Flagstaff Area National Monuments.

Holm, R. F. 1986. Field guide to the geology of the central San Francisco Volcanic Field, northern Arizona. In Geology of central and northern Arizona, edited by J. D. Nations, C. M. Conway, and G. A Swann, pp. 27–41. Geological Society of America, Rocky Mountan Section Guidebook. Flagstaff.

Hooten, J. A., M. H. Ort, and M. D. Elson 2001. The origin of cinders in Wupatki National Monument. Report prepared for Southwest Parks and Monuments Association Grant FY00-11. Technical Report 2001-12. Desert Archaeology, Tucson AZ.

Houghton, J. T., Y. Ding, D. J. Griggs, M. Noguer, P. J. van der Linden, and D. Xiasou, Editors. 2001. Climate change 2001: The scientific basis. Cambridge University Press, Cambridge UK.

Hughes, M. K., and H. F. Diaz. 1994. Was there a Medieval warm period, and if so, where and when? Climatic change 26: 109–142.

IPCC, Intergovernmental Panel on Climate Change. 1998. The terrestrial carbon cycle: Implications for the Kyoto Protocol. Science 280: 1393–1394.

IPCC, Intergovernmental Panel on Climate Change. 2000. Land use, land-use change, and forestry. Cambridge University Press, Cambridge UK.

IPCC, Intergovernmental Panel on Climate Change. 2001. Climate change 2001: Impacts, adaptations, and vulnerability. Contribution of working group II to the third assessment report of the IPCC, edited by J. J. McCarthy, O.F. Canziani, N.A. Leary, D. J. Dokken, and K. S. White. Cambridge University Press, Cambridge UK.

Irwin, L. L., J. G. Cook, R. A. Riggs, and J. M. Skovlin. 1993. Effects of long-term grazing by big game and livestock in the Blue Mountains forest ecosystems. In Eastside forest ecosystem health assessment, vol. 3: Assessment, compiled by P. F. Hessburg, pp. 537–588. U.S. Department of Agriculture, Forest Service, Pacific Northwest Research Station, Portland OR.

Jackson, R. B., J. L. Banner, E. G. Jobbagy, W. T. Pockmand, and D. H. Wall. 2002. Ecosystem carbon loss with woody plant invasion of grasslands. Nature (418): 623–626.

Jacobs, B. F. 1989. A flora of Bandelier National Monument. Unpublished report on file at Bandelier National Monument, NM.

Jacobs, B. F. 2002. Vegetation management plan: Bandelier National Monument. Manuscript on file, Bandelier National Monument, NM.

Jacobs, B. F., R. G. Gatewood, and C. D. Allen. 2002. Watershed restoration in degraded piñon-juniper woodlands: A paired watershed study 1996–98(9). Unpublished report on file at Bandelier National Monument, NM.

Jenkins, M. 2001. Making forests safe again won't be a walk in the park. High Country News. 7 May, p. 10.

Jones, E. T., R. J. McLain, and J. Weigand. 2002. Non-timber forest products in the United States. University of Kansas Press, Lawrence.

Jones, E. T., R. J. McLain, and K. A. Lynch. 2004. The relationship between non-timber forest product management and biodiversity in the United States. Report submitted to the National Commission of Science for Sustainable Forestry. Institute for Culture and Ecology, Portland OR. (www.ifcae.org/projects/ncssf1).

Julius, C. 1999. A comparison of vegetation structure on three different soils at Bandelier National Monument, New Mexico. Master's thesis, Rheinischen Friedrich-Wilhelms University, Bonn, Germany.

Kaib, J. M. 1998. Fire history in riparian canyon pine-oak forests and intervening desert grasslands of the Southwest borderlands: A dendrochronological, historical and cultural inquiry. Master's thesis, University of Arizona, Tucson.

Kaib, J. M., C. H. Baisan, H. D. Grisson-Mayer, and T. W. Swetnam. 1996. Fire history in the gallery pine-oak forests and adjacent grasslands of the Chiricahua Mountains of Arizona. In Effects of fire on Madrean Province ecosystems, edited by P. Ffolliott et al., pp. 253–264. USDA.

Kloor, K. 2000. Returning America's forests to their "natural" roots. Science 287: 573–575.

Kohler, T. A., and M. H. Matthews. 1988. Long-term Anasazi land use and forest reduction: A case study from southwest Colorado. American Antiquity 53(3): 537–564.

Krech, S. 1999. The ecological Indian. W. W. Norton, New York.

Kuske, C. R., L. O. Ticknor, J. D. Busch, C. A. Gehring, and T. G. Whitham. 2003. The pinyon rhizosphere, plant stress, and herbivory affect the abundance of microbial decomposers in soils. Microbial Ecology 45: 340–352.

Langston, N. 1999. Environmental history and restoration in the western forests. Journal of the West 38(4):45-56.

Leatherman, D. A., and B. C. Kondratieff. 2003. Insects associated with the pinyon-juniper woodlands of Mesa Verde country. In Ancient pinyon-juniper woodlands: A natural history of Mesa Verde country, edited by M. L. Floyd, pp. 167–182. University Press of Colorado, Boulder.

Leopold, A. 1943. Deer irruptions. Wisconsin Conservation Bulletin 8: 3–11.

Lipe, W. D., and B. L. Pitblado. 1999. Paleoindian and Archaic periods. In Colorado prehistory: A context for the southern Colorado River basin, edited by W. D. Lipe, M. D. Varien, and R. H. Wilshusen, pp. 95–131. Colorado Council of Professional Archaeologists, Denver.

Lipe, W. D., and M. D. Varien. 1999a. Pueblo II (A.D. 900–1150). In Colorado prehistory: A context for the southern Colorado River basin, edited by W. D. Lipe, M. D. Varien, and R. H. Wilshusen, pp. 242–289. Colorado Council of Professional Archaeologists, Denver.

Lipe, W. D., and M. D. Varien. 1999b. Pueblo III (A.D. 1150–1300). In Colorado prehistory: A context for the southern Colorado River basin, edited by W. D. Lipe, M. D. Varien, and R. H. Wilshusen, pp. 290–352. Colorado Council of Professional Archaeologists, Denver.

Little, E. L., Jr. 1971. Atlas of United States trees, vol. 1. USDA Forest Service Miscellaneous Publications 1146, Washington D.C.

Loeser, M. R., T. D. Sisk, and T. E. Crews. 2001. Plant community responses to livestock grazing: An assessment of alternative management practices in a semiarid grassland. USDA Forest Service Proceedings RMRS-P-22.

Lopinot, N. H. 1986. The early Spanish introduction of new cultigens into the greater Southwest. Missouri Archaeologist 47: 61–84.

Lynch, A. M., and T. W. Swetnam. 2003. Southwestern climate trends and forest insects: Small changes with amplified responses. Abstract, In Seventh Biennial Conference of Research on the Colorado Plateau. Program and Abstracts of Presented Papers and Posters. November 2003.

Mac, M. J. 1998. Status and trends of the nation's biological resources. U.S. Geological Survey, Reston VA.

Madany, M. H., and N. E. West. 1983. Livestock grazing–fire regime interactions within montane forests of Zion National Park, Utah. Ecology 64(4): 661–667.

Mann, M. E., R. S. Bradley, and M. K. Hughes. 1999. Northern Hemisphere temperatures during the past millennium: Inferences, uncertainties, and limitations. Geophysical Research Letters 26(6): 759–762.

Massey, C. L., D. D. Lucht, and J. M. Schmid. 1977. Roundheaded pine beetle. Forest Insect and Disease Leaflet 155. USDA Forest Service, Washington DC.

Matthews, M. H. 1987. The Dolores Archaeological Program macrobotanical data base: Resource availability and mix. In Dolores Archaeological Program: Final synthetic report, edited by D. A. Breternitz, C. K. Robinson, and G. T. Gross, pp. 199–207. U.S. Department of Interior, Denver.

McCullough, D., R. A. Werner, and D. Neumann. 1998. Fire and insects in northern and boreal forest ecosystems of North America. Annual Review Entomology 43: 107–122.

McFadden, L. D., and J. R. McAuliffe. 1997. Lithologically influenced geomorphic responses to Holocene climatic changes in the southern Colorado Plateau, Arizona: A soil-geomorphic and ecologic perspective. Geomorphology 19: 303–332.

McLaren, B. E., and R. O. Peterson. 1994. Deer irruptions. Wisconsin Conservation Bulletin 8: 3–11.

McVickar, J. 1999. Analysis of plant macrofossils. In Anasazi community development in Cove–Redrock Valley. Archaeological excavations along the N33 road in Apache County, Arizona, Vol. II, edited by P. F. Reed and K. N. Hensler, pp. 809–850. Navajo Nation Papers in Anthropology 33. Window Rock AZ.

Meinecke, E. P. 1935. Observations in Mesa Verde National Park in 1935. Manuscript on file, Mesa Verde National Park, CO.

Milankovitch, M. M. 1941. Canon of insolation and the ice-age problem. Beograd: Königlich Serbische Akademie. English translation by the Israel program for scientific translations, published for the U.S. Department of Commerce and the National Science Foundation, Washington DC.

Millar, C. I. and W.B. Woolfenden. 1999. The role of climate change in interpreting historical variability. Ecological Applications 9: 1207–1216.

Miller, R. F., and R. J. Tausch. 2001. The role of fire in pinyon and juniper woodlands: A descriptive analysis. In Proceedings of the Invasive Species Workshop: The Role of Fire in the Control and Spread of Invasive Species, edited by K. E. M. Galley and T. P. Wilson, pp. 15–30. Fire Conference 2000: The First National Congress on Fire Ecology, Prevention, and Mangement. Miscellaneous Publication 11, Tall Timbers Research Station, Tallahasee FL.

Mitchell, J. E., and T. C. Roberts, Jr. 1999. Distribution of pinyon-juniper in the western United States. In Ecology and management of pinyon-juniper communities within the interior west, edited by S. B. Monsen and R. Stevens, pp. 146–154. USDA Forest Service Proceedings RMRS-P-9, Rocky Mountain Research Station, Fort Collins.

Mittlebach, G. C., A. M. Turner, D. J. Hall, J. E. Retting, and C. W. Osenberg. 1995. Perturbation and resilience: A long-term, whole lake study of predator extinction and reintroduction. Ecology 76: 2347–2360.

Moerman, D. E. 1998. Native American ethnobotany. Timber Press, Portland OR.

Monti, L. 2003. Documenting and incorporating Navajo traditional land management and ecological restoration practices into pine and fir uplands of the Asayi watershed of the Chuska Mountains. Final report to the Ecological Restoration Institute's Southwest Fire Initiative. Center for Sustainable Environments, Northern Arizona University, Flagstaff.

Moore, R. B. 1974. Geology, petrology and geochemistry of the eastern San Francisco Volcanic Field, Arizona. Ph.D. dissertation, Dept. of Geosciences, University of New Mexico, Albuquerque.

Moore, M. M., and D. Huffman. In press. Tree encroachment into meadows on the Kaibab Plateau. Arctic, Antarctic and Alpine Research.

Moote, A., and D. Becker. 2003. Exploring barriers to collaborative forestry. Ecological Restoration Institute (SAF and ICAF), Flagstaff AZ.

Mopper, S., J. Maschinski, N. Cobb, and T. G. Whitham. 1991a. A new look at habitat structure: Consequences of herbivore-modified plant architecture. In Habitat structure: The physical arrangements of objects in space, edited by S. S. Bell, E. D. McCoy, and H. R. Mushinsky. Chapman and Hall, London.

Mopper, S., J. B. Mitton, T. G. Whitham, N. S. Cobb, and K. M. Christensen. 1991b. Genetic differentiation and heterozygosity in pinyon pine associated with resistance to herbivory and environmental stress. Evolution 45(4): 989–999.

Nabhan, G. P. 1998. Cultural parallax in viewing North American habitats. In The great wilderness debate, edited by J. B. Callicott and M. P. Nelson, pp. 628–534. University of Georgia Press, Athens.

Nabhan, G. P. 2003. Assessing Arizona's food security, safety and sustainability: Weighing the risks. Center for Sustainable Environments White Paper, Northern Arizona University, Flagstaff. 8 pp.

Nabhan, G. P. 2004. Destruction of an indigenous cultural landscape: An epitaph from Organ Pipe National Monument. Ecological Restoration 21(4): 290–295.

Nabhan, G. P., and M. K. Anderson. 1991. Gardeners in Eden. Wilderness Magazine 55(194): 27–34.

Nabhan, G. P., and C. C. Wilson. 1995. Canyons of color. Harper-Collins West, San Francisco.

Nabhan, G. P., P. Pynes, and T. Joe. 2002a. Safeguarding the uniqueness of the Colorado Plateau. Northern Arizona Univrsity Center for Sustainable Environments, Terralingua, and Grand Canyon Wildlands Council, Flagstaff AZ.

Nabhan, G. P., P. Pynes, and T. Joe. 2002b. Assessing levels of biocultural diversity on the Colorado Plateau in relation to other regions. In Safeguarding the uniqueness of the Colorado Plateau: An ecoregional assessment of biocultural diversity. Northern Arizona University Center for Sustainable Environments, Flagstaff AZ.

Nabhan, G. P., P. Pynes, and T. Joe. 2002c. Safeguarding species, languages and cultures in the time of diversity loss: From the Colorado Plateau to global hotspots. Annals of the Missouri Botanical Garden 89(2): 164–175.

National Oceanic and Atmospheric Administration. 2004. Abrupt climate change. http://www.ncdc.noaa.gov/paleo/abrupt/data_glacial.html

National Park Service. 1942. Chief Counsel, U.S. Department of Interior, National Park Service, August 24, 1942, to Superintendent Nussbaum, Mesa Verde National Park. Memorandum on file, Mesa Verde National Park.

National Park Service. 2000. BAER Plan. Chapin #5 fire, burned areas emergency rehabilitation plan. Ms on file, Mesa Verde National Park.

National Park Service. 2002a. Sunset Crater final environmental impact statement, general management plan. U.S. Department of the Interior, National Park Service, Washington DC.

National Park Service. 2002b. Wupatki final environmental impact statement, general management plan. U.S. Department of the Interior, National Park Service, Washington DC.

National Park Service. 2003. Annual performance plan for Flagstaff area national monuments: Wupatki, Sunset Crater, and Walnut Canyon National Monuments, Fiscal Year 2003. Administrative files, Flagstaff Area National Monuments. http://www.nps.gov/sucr/index.htm.

National Research Council (U.S.) Committee on Abrupt Climate Change. 2002. Abrupt climate change: Inevitable surprises. Committee on Abrupt Climate Change, Ocean Studies Board, Polar Research Board on Atmospheric Sciences and Climate, Division on Earth and Life Studies, National Research Council. National Academy Press, Washington DC.

Nature Conservancy. 1996. America's least wanted: Alien species invasions of U.S. ecosystems. The Nature Conservancy, Arlington VA.

Neftel, A., H. Friedli, E. Moor, H. Lötscher, H. Oeschger, U. Siegenthaler, and B. Stauffer. 1994. Historical CO2 record from the Siple Station ice core. In Trends: A compendium of data on global change. Carbon Dioxide Information Analysis Center, Oak Ridge National Laboratory, U.S. Department of Energy, Oak Ridge TN. http://cdiac.esd.ornl.gov/trends/co2/siple.htm

Nijhuis, M. 1999. Flagstaff searches for its forests' future. High Country News, 1 March, pp. 8–12.

Nijhuis, M. 2004. Attack of the bark beetles. High Country News 36(13): 8–14.

Northern and Southern Colorado Plateau Networks. 2004. Colorado Plateau (COPL) exotic plant management team proposal: FY 2002 through FY 2006. Manuscript on file, Cooperative Ecosystems Studies Unit, Northern Arizona University, Flagstaff.

Ogle, K., T. G. Whitham, and N. Cobb. 2000. Tree-ring variation in pinyon predicts likelihood of death following severe drought. Ecology 81(11): 3232–3243.

Omi, P. N., and R. L. Emrick. 1980. Fire and resource management in Mesa Verde National Park. Manuscript on file, Mesa Verde National Park, CO.

Orodho, A. B., M. J. Trlica, and C. D. Bonham. 1990. Long-term heavy grazing effects on soil and vegetation in the Four Corners region. Southwestern Naturalist 35: 9–14.

O'Rourke, P. M. 1980. Frontier in transition: A history of southwest Colorado. Bureau of Land Management Colorado State Office, Denver.

Ort, M. H, M. D. Elson, and D. E. Champion. 2002. A paleomagnetic dating study of Sunset Crater Volcano. Investigator's final report SUCR-00007. Technical Report 2002-16. Desert Archaeology, Inc., Tucson.

Overpeck, J. T. 1996. Warm climate surprises. Science 271 (5257): 1820–1822.

Parmenter, R. R., and T. R. Van Devender. 1995. Diversity, spatial variation, and functional roles of vertebrates in the desert grassland. In Desert grasslands, edited by M. P. McClaran and T. R. Van Devender, pp. 196–229. University of Arizona Press, Tucson.

Petersen, K. L. 1988. Climate and the Dolores River Anasazi. University of Utah Anthropological Paper 113, Salt Lake City.

Petersen, K. L. 1994. A warm and wet Little Climatic Optimum and a cold and dry Little Ice Age in the southern Rocky Mountains, U.S.A. Climatic Change 26: 243–269.

Petersen, K. L., and P. J. Mehringer, Jr. 1976. Postglacial timberline fluctuations, La Plata Mountains, southwestern Colorado. Arctic and Alpine Research 8(3): 275–288.

Petersen, K. L. 1986. Pollen studies: Temporal patterns in resource use. In Dolores Archaeological Program: Final synthetic report, edited by D. A. Breternitz, C. K. Robinson, and G. T. Gross, pp. 184–199. U.S. Department of the Interior, Denver.

Pilles, P. J., Jr. 1993. The Sinagua: Ancient people of the Flagstaff region. In Wupatki and Walnut Canyon: New perspectives on history, prehistory and rock art, edited by D. G. Noble, pp. 2–11. Ancient City Press, Santa Fe NM.

Pilsk, S., and J. Cassa. 2004. The Western Apache home: Landscape management and failing ecosystems. Paper presented at Biodiversity and Management of the Madrean Archipelago II: Connecting Mountain Islands and Desert Seas, May 11–14, 2004. Tucson AZ.

Pimental, D., L. Lach, R. Zuniga, and D. Morrison. 2000. Environmental and economic costs of non-indigenous species in the United States. Bioscience 50: 53–66.

Powell, J. W. 1890. The non-irrigible lands of the arid region. Century Magazine 39: 915–922.

Price, C., and D. Rind. 1994. The impact of 2 x CO2 climate on lightning-caused fires. Journal of Climatology 7: 1484–1494.

Preisler, H. K., and R. G. Mitchell. 1993. Colonization patterns of the mountain pine beetle in thinned and unthinned lodgepole pine stands. Forest Science 39(3): 528–545.

Priest, S. S., W. A. Duffield, K. Malis-Clark, J. W. Hendley, II, and P. H. Stauffer. 2001. The San Francisco Volcanic Field, Arizona. U.S. Geological Survey Fact Sheet 017-01.

Prouty, G. 1997. Paleoethnobotanical investigations. In Archaeological nature and extent testing, HAER documentation of one bridge site, and identification and assessment of a Chacoan linear feature along Navajo Route N5m001(1), Toadlena to Newcomb, San Juan County, New Mexico, pp. 467–489. Zuni Cultural Resource Enterprise Report 512, Zuni NM.

Pyne, S. J. 2001. Year of the fires: The story of the great fires of 1910. Penguin Books, New York.

Pynes, P. G. 2000. Erosion, extraction, reciprocation: An ethno/environmental history of the Navajo Nation's ponderosa pine forests. Ph.D. dissertation, University of New Mexico, Albuquerque.

Rampino, M. R., and S. Self. 1984. Sulphur-rich volcanic eruptions and stratosphere aerosols. Nature 310: 677–679.

Ramsey, D. 2003. Soils of Mesa Verde Country. In Ancient pinyon-juniper woodlands: A natural history of Mesa Verde country, edited by M. L. Floyd, pp. 213–222. University of Colorado Press, Boulder.

Reed, P., and K. Hensler. 1999. Anasazi community development in Cove–Red Rock Valley. Navajo Nation Papers in Anthropology 33. Navajo Nation Archaeology Division. Window Rock AZ.

Reneau, S. L., and E. V. McDonald. 1996. Landscape history and processes on the Pajarito Plateau, northern New Mexico: Rocky Mountain cell. Friends of the Pleistocene, Field Trip Guidebook. Los Alamos NM.

Ricketts, T. H., E. Dinerstein, D. M. Olson, and C. Loucks. 1999a. Who's where in North America? BioScience 49(5): 369–381.

Ricketts, T. H., E. Dinerstein, D. M. Olson, and C. Loucks. 1999b. Terrestrial ecoregions of North America: A conservation assessment. Island Press, Washington DC.

Rink, G. 2003. Floristic inventory of Canyon de Chelly National Monument. Master's thesis, Northern Arizona University, Flagstaff.

Roberts, A. 1990. Navajo ethno-history and archaeology. In The Wupatki Archaeological Inventory Survey Project: Final report, edited by B. A. Anderson, Southwest Cultural Resources Professional Paper 35. Santa Fe NM.

Rogers, P. 2002. Mesa Verde fire history. Manuscript on file, Mesa Verde National Park, CO.

Romme, W. H., S. Oliva, and M. L. Floyd. 2003a. Threats to the pinyon-juniper woodlands. In Ancient pinyon-juniper woodlands: A natural history of Mesa Verde country, edited by M. L. Floyd, pp. 339–360. University of Colorado Press, Boulder.

Romme, W. H., R. Balice, P. Brown, N. S. Cobb, T. DeGomez, L. Floyd-Hanna, P. Fulé, D. W. Huffman, B. A. Hungate, G. W. Koch, M. M. Moore, M. Savage, and E. W. Schupp. 2003b. Letter to U.S. Department of Agriculture, http://www.mpcer.nau.edu/direnet/files/Romme2003.pdf.

Romme, W. H., L. Floyd-Hanna, and D. D. Hanna. 2003c. Ancient pinyon-juniper forests of Mesa Verde and the West: A cautionary note for restoration programs. In Fire, fuel treatments, and ecological restoration, edited by P. N. Omi and L. A. Joyce, pp. 335–350. USDA Forest Service Proceedings RMRS-P-29, Rocky Mountain Research Station, Fort Collins CO.

Rothman, H. K. 1992. On rims and ridges: The Los Alamos area since 1880. University of Nebraska Press, Lincoln.

Ruel, J., and T. G. Whitham. 2002. Fast-growing juvenile pinyons suffer greater herbivory when mature. Ecology 83(10): 2691–2699.

Rundall, J. M., M. Z. Mier, and N. S. Cobb. 2003. Documenting past pinyon-juniper woodland treatments on Colorado Plateau Bureau of Land Management (BLM) lands. Poster presented at Seventh Biennial Conference of Research on the Colorado Plateau, Northern Arizona University, Flagstaff.

Salzer, M. 2000a. Dendroclimatology in the San Francisco Peaks region of North America, USA. Ph.D. dissertation, University of Arizona, Tucson.

Salzer, M. 2000b. Temperature variability and the northern Anasazi: Possible implications for regional abandonment. Kiva 65(4): 295–318.

Salzer, M. W., and J. S. Dean. 2004. Dendroclimatic reconstructions and paleoenvironmental analyses. In Sunset Crater archaeology: The history of a volcanic landscape, edited by M. D. Elson. Anthropological Papers 37. Center for Desert Archaeology, Tucson.

Sampson, A. W. 1944. Plant succession on burned chaparral lands in northern California. Bull. 65. Berkeley, CA: University of California, College of Agriculture, Agricultural Experiment Station. 144 p.

San Miguel, G. L., and M. Colyer. 2003. Mesa Verde country's woodland avian community. In Ancient pinyon-juniper woodlands: A natural history of Mesa Verde country, edited by M. L. Floyd, pp. 89–110. University of Colorado Press, Boulder.

Savage, M. 1989. Structural dynamics of a pine forest in the American Southwest under chronic human disturbance. Ph.D. dissertation, University of Colorado, Boulder.

Savage, M. 2003. Restoring natural systems through natural processes. Quivira Coalition Newsletter 6(2): 1–27.

Savage, M., and T. W. Swetnam. 1990. Early nineteenth century fire decline following sheep pasturing in a Navajo ponderosa pine forest. Ecology 71(6): 2374–2378.

Schneider, S. 2004. Global warming: Neglecting the complexities. Scientific American 286(1): 62–65.

Schroeder, A. H. 1977. Of men and volcanoes: The Sinagua of northern Arizona. Southwest Parks and Monuments Association, Globe AZ.

Selmants, P. C., A. Elseroad, and S. C. Hart. 2003. Soils and nutrients. In Peter Friederici, editor. In Ecological restoration of southwestern ponderosa pine forests, edited by P. Friederici, pp. 144–160. Island Press, Washington DC.

Shoemaker, E. M., and D. E. Champion. 1977. Eruption history of Sunset Crater, Arizona. Investigator's Annual Report. Manuscript on file, Flagstaff Area National Monuments Headquarters.

Short, M. S. 1988. Walnut Canyon and Wupatki: A history. Master's thesis, Northern Arizona University, Flagstaff.

Smith, D. A. 2003. Only man is vile. In Ancient pinyon-juniper woodlands: A natural history of Mesa Verde country, edited by M. L. Floyd, pp. 321–336. University of Colorado Press, Boulder.

Smith, R. L., R. A. Bailey, and C. S. Ross. 1970. Geologic map of the Jemez Mountains, New Mexico. Miscellaneous Geologic Investigations Map I-571. U.S. Geologic Survey, Washington DC.

Smith, S. 1997. Pollen analysis. In Archaeological nature and extent testing, HAER documentation of one bridge site, and identification and assessment of a Chacoan linear feature along Navajo Route N5001(1), Toadlena to Newcomb, San Juan County, New Mexico, pp. 417–426. Zuni Cultural Resource Enterprise Report 512, Zuni NM.

Smith, S. 1998. Pollen analysis. In Fire in the hole: The effects of fire on subsurface archaeological materials, edited by Ruscavage-Barz et al., pp. 155–170. Unpublished manuscript on file at Bandelier National Monument.

Smith, S. 1999. Pollen analysis. In Anasazi community development in Cove–Redrock Valley. Archaeological excavations along the N33 road in Apache County, Arizona, Vol. II, edited by P. F. Reed and K. N. Hensler, pp. 851–869. Navajo Nation Papers in Anthropology 33. Window Rock AZ.

Smith. S. J. 2004. The U.S. 89 pollen analysis and regional archaeobotanical overview. In Sunset Crater archaeology: The history of a volcanic landscape. Environmental analyses (draft), edited by M. D. Elson. Anthropological Papers 33, Center for Desert Archaeology, Tucson.

Smith, S., M. Matthews, K. N. Hensler, and L. A. Schniebs. 1999. Analyses of subsistence: Pollen, macrobotanical, and faunal results. In A Pueblo household on the Chuska slope. Data recovery at NM-H-47-102, along Navajo Route 5010(1) near Toadlena, New Mexico. Navajo Nation Papers in Anthropology 35. Window Rock AZ.

Snider, G. B., D. B. Wood, and P. J. Daugherty. 2003. Analysis of costs and benefits of restoration-based hazardous fuel reduction treatments versus no treatment. School of Forestry Progress Report, Northern Arizona University, Flagstaff.

Soulé, M. E., J. A. Estes, J. Berger, and C. M. Del Rio. 2003. Ecological effectiveness: Conservation goals for interactive species. Conservation Biology 17(5): 1238–1251.

Southwest Forest Alliance. 2000. Forest restoration: Natural processes restoration model. SFA, Flagstaff AZ.

Stanislawski, M. B. 1963. Wupatki pueblo: A study in cultural fusion and change in Sinagua and Hopi prehistory. Ph.D. dissertation, University of Arizona, Tucson.

Stebbins, G. I. 1981. Coevolution of grasses and herbivores. Annals of the Missouri Botanical Garden 68: 75–86.

Stiger, M. A. 1979. Mesa Verde subsistence patterns from Basketmaker to Pueblo III. The Kiva 44(2–3): 133–144.

Superintendent's Annual Report. 1946. Mesa Verde National Park. Manuscript on file, Mesa Verde National Park.

Swaty, R. L. R. J. Deckert, T. G. Whitham, and C. A. Gehring. In press. Ectomycorrihzal abundance and community composition shifts with drought: Predictions from tree rings.

Swetnam, T. W., and C. H. Baisan. 1996. Historical fire regime patterns in the southwestern United States since A.D. 1700. In Fire effects in southwestern forests, edited by C. D. Allen, pp. 11–32. USDA Forest Service General Technical Report RM-286, Rocky Mountain Research Station, Fort Collins CO.

Swetnam, T. W., and C. H. Baisan. 2003. Tree-ring reconstructions of fire and climate history in the Sierra Nevada and southwestern United States. In fire and climatic change in temperate ecosystems of the Western Americas, edited by T. T. Veblen, W. Baker, G. Montenegro, and T. W. Swetnam, pp. 158–105. Springer, New York.

Swetnam, T. W., and J. L. Betancourt. 1990. Fire–Southern Oscillation relations in the southwestern United States. Science 249: 1017–1021.

Swetnam, T. W., and A. M. Lynch. 1993. Multicentury, regional-scale patterns of western spruce budworm outbreaks. Ecological Monographs 63(4): 399–424.

Swetnam, T. W., C. D. Allen, and J. L. Betancourt. 1999. Applied historical ecology: Using the past to manage for the future. Ecological Applications 9: 1189–1206.

Tausch, R. J., C. L. Nowak, and S. Mensing. 2004. Potential implications of Holocene climate variation for the extent, severity and duration of the current drought. Seventh Biennial Conference of Research on the Colorado Plateau. Northern Arizona University, Flagstaff.

Terborgh, J. 2001. Ecological meltdown in predator-free forest fragments. Science 294: 1923–1925.

Terborgh, J., and B. Winter. 1980. Some causes of extinction. In Conservation biology: An evolutionary-ecological perspective, edited by M. E. Soulé and B. A. Wilcox. Sinauer, Sunderland MA.

Thomas, L., J. Whittier, N. Tancreto, J. Atkins, M. Miller, and A. Cully. 2003. Vital signs monitoring plan for the southern Colorado Plateau Network: Phase I report. Final draft September 29, 2003. Manuscript on file, Cooperative Ecosystems Studies Unit, Northern Arizona University, Flagstaff.

Thompson, R. S., C. Whitlock, P. J. Bartlein, S. P. Harrison, and W. G. Spaulding. 1993. Climatic changes in the western United States since 18,000 yr BP. In Global climates since the last glacial maximum, edited by H. E. Wright, J. E. Kutzbach, Jr., T. Webb, III, W. F. Ruddiman, F. S. Street-Perrott, and P. J. Bartlein. University of Minnesota Press, London.

Tiedmann, A. R., W. P. Clary, and R. J. Barbour. 1987. Underground systems of gambel oak (Quercus gambelii) in central Utah. American Journal of Botany. 74(7): 1065-1071.

Toll, H. W., Editor. 1995. Soil, water, biology, and belief in prehistoric and traditional southwestern agriculture. New Mexico Archaeological Council, Special Publication 2.

Toll, M. S., and P. J. McBride. 1996. Botanical contents of two 17th c. baskets from a dry shelter in the Galisteo Basin, NM. Museum of of New Mexico, Office of Archaeological Studies, Technical Series 42.

Torres-Reyes, R. 1970. Mesa Verde National Park: An administrative history 1906–1970. U.S. Department of the Interior, Office of History and Historic Architecture, Eastern Service Center, Washington DC.

Touchan, R, T. W. Swetnam, and H. D. Grisson-Mayer. 1995. Effects of livestock grazing on pre-settlement fire regimes in New Mexico. In Proceedings: Symposium on fire in wilderness and park management, edited by J. K. Brown, R. W. Mutch, C. W. Spoon, and R. H. Wakimoto, pp. 268–272. General Technical Report INT-GTR-320.USDA Forest Service, Ogden UT.

Touchan, R., C. D. Allen, and T. W. Swetnam. 1996. Fire history and climatic patterns in ponderosa pine and mixed-conifer forests of the Jemez Mountains, northern New Mexico. In Proceedings of the 1994 Symposium on the La Mesa Fire, edited by C. D. Allen, pp. 33–46. USDA Forest Service General Technical Report RM-286, Fort Collins CO.

Travis, S. E. 1990. Chapter 4. In The Wupatki Archaeological Inventory Survey Project: Final report, edited by B. Anderson, pp. 1–38. Southwest Cultural Resources Professional Paper 35. Santa Fe NM.

Trewartha, G. T. 1954. An introduction to climate. McGraw-Hill, New York.

Trimble, M. 1982. CO Bar. Bill Owen Depicts the Historic Babbitt Ranch. Northland Press, Flagstaff. 95 pp.

Trotter, R. T. III, N. S. Cobb, and T. G. Whitham. 2002. Herbivory, plant resistance, and climate in the tree ring record: Interactions distort climatic reconstructions. PNAS 99(15).

Trotter, R. T. III, N. S. Cobb, and T. G. Whitham. 2004. Arthropod community diversity and trophic structure respond to plant stress: Implications for global change. Oecologica, in press.

Underhill, R. M. 1971. The Navajos. University of Oklahoma Press, Norman.

Union of Concerned Scientists. 2004. Scientific integrity in policymaking: An investigation into the Bush administration's misuse of science. Union of Concerned Scientists, Cambridge MA.

University of Arizona Cooperative Extension. 2004. Available on Arizona Health Web page: http://ag.arizona.edu/extension/fh/insects.html

U.S. Department of Agriculture. 2001. National strategy for special forest products. Washington DC.

U.S. Department of Agriculture, National Forest Service. 2002. Forest insect and disease conditions in the southwestern region, 2002. http://www.fs.fed.us/r3/publications/documents/fidc2002.pdf.

U.S. General Accounting Office. 1999. Western national forests: A cohesive strategy is needed to address catastrophic fire threats. U.S. Government Printing Office GAO/RCED-99-65, Washington DC. http:///www.gao.gov.

Van Hooser, D. D., R. A. O'Brien, and D. C. Collins. 1993. New Mexico's forest resources. USDA Forest Service Bulletin INT-79, Albuquerque NM.

Van West, C. R., and J. S. Dean. 2000. Environmental characteristics of the A.D. 900–1300 period in the central Mesa Verde region. Kiva 66(1): 19–44.

Vierra, B., S. R. Hoagland, J. S. Isaacson, and A. L. Madsen. 2002. Department of Energy land conveyance data recovery plan and research design for the excavation of archaeological sites located within selected parcels to be conveyed to the incorporated County of Los Alamos, New Mexico. Los Alamos National Laboratory LA-UR-02-1284.

Vitousek, P. M. 1990. Biological invasions and ecosystem process: Towards an integration of population biology and ecosystem studies. Oikos 57: 7–13.

Wagner, A. 2004. Western drought beats Dust Bowl, could be worst in 500 years. June 17. Associated Press.

Wagner, F. H. 1978. Livestock grazing and the livestock industry. In Wildlife and America, edited by H. P. Brokaw, pp. 121–145. Council on Environmental Quality, Washington DC.

Wallace, T. R. 1949. A brief history of Coconino County. Master's thesis. Arizona State College (Northern Arizona University), Flagstaff.

Weng, C., and S. T. Jackson. 1999. Late-glacial and Holocene vegetation history and paleoclimate of the Kaibab Plateau, Arizona. Palaeogeography, Palaeoclimatology, and Palaeoecology 153: 179–201.

Westoby, M., B. Walker, and I. Noy-Meir. 1989. Opportunistic management for rangelands not at equilibrium. Journal of Range Management 42: 266–274.

Whitham, T. G., and S. Mopper. 1985. Chronic herbivory: Impacts on tree architecture and sex expression of pinyon pine. Science 227: 1089–1091.

Wilcove, D. S., D. Rothstein, J. Dubow, A. Phillips, and E. Losos. 1998. Quantifying threats to imperiled species in the United States. BioScience 48: 607–615.

Wilmers, C. C., R. L. Crabtree, D. W. Smith, K. M. Murphy, and W. M. Getz. 2003. Trophic facilitation by introduced top predators: Grey wolf subsidies to scavengers in Yellowstone National Park. Journal of Animal Ecology 72: 909–916.

Wilshusen, R. H. 1999. Pueblo I (A.D. 750–900). In Colorado prehistory: A context for the southern Colorado River basin, edited by W. D. Lipe, M. D. Varien, and R. H. Wilshusen, pp. 196–241. Colorado Council of Professional Archaeologists, Denver.

Wilshusen, R. H., and R. H. Towner. 1999. Post-Puebloan occupation (A.D. 1300–1840). In Colorado prehistory: A context for the southern Colorado River basin, edited by W. D. Lipe, M. D. Varien, and R. H. Wilshusen. Colorado Council of Professional Archaeologists, Denver.

Wood, R. E. 1983. Mortality caused by root diseases and associated pests on six national forests in Arizona and New Mexico. Forest Pest Management Report R3-83-13. USDA Forest Service, Southwest Region, Albuquerque NM.

Woodhead, P. V. 1946. Cooperative range training trip, Region 2 and Mesa Verde National Park. Manuscript on file, Mesa Verde National Park, CO.

Wright, H. A., L. F. Newenschwander, and C. M. Britton. 1979. The role and use of fire in sagebrush-grass and pinyon-juniper plant communities: A state-of-the-art review. U.S. Department of Agriculture, Forest Service General Technical Report INT-58. Intermountain Forest and Range Experiment Station, Ogden UT.

Wright, H. E., Jr., A. M. Bent, B. S. Hansen, and L. J. Maher, Jr. 1973. Present and past vegetation of the Chuska Mountains, northwestern New Mexico. Geological Society of America Bulletin 84: 1155–1180.

Wycoff, D. G. 1977. Secondary forest succession following abandonment of Mesa Verde. The Kiva 42(3–4): 215–231.

Zavaleta, E. 2000. Valuing ecosystem services lost to *Tamarix* invasion in the United States. In H.A. Mooney and R. Hobbs, editors, In Invasive species in a changing world, edited by H. A. Mooney and R. Hobbs, pp. 261–300. Island Press, Washington DC.

Appendix A

INVASIVE PLANT SPECIES FROM OUR STUDY AREAS

Appendix A. Invasive plant species from our study areas; asterisk denotes native.

Family	Scientific Name	Common Name	Jemez	Mesa Verde	Wupatki	Chuskas
Amaranthaceae	Amaranthus albus*	White amaranth, pigweed	x	–	x	–
	A. hybridus*	Amaranth	x	–	–	–
	A. hypochondriacus	Love-lies-bleeding	x	–	–	–
	A. palmeri*	Careless weed	x	–	–	x
	A. retroflexus	Redroot pigweed	x	–	x	–
Apiaceae	Cicuta douglasii*	Hemlock	–	–	–	x
	Conium maculatum	Poison hemlock	x	–	–	–
Apocynaceae	Apocynum cannabinum*	Indian hemp	x	–	–	–
Asclepidaceae	Asclepias suoberticillata*	Poison milkweed	x	–	x	x
	A. tuberosa*	Butterfly weed	x	–	–	–
Asteraceae	Acroptilon repens	Hardheads	x	x	–	x
	Bidens bipinnata*	Beggartick, Spanish needles	x	–	–	–
	B. cernua*	Nodding beggartick	x	–	–	–
	B. frondosa*	Sticktight	x	–	–	–
	B. tripartita*	Beggartick	x	–	–	–
	Carduus nutans	Musk thistle	x	x	–	x
	Centaurea dehisa	Knapweed	–	x	–	x
	C. solstitialis	Yellow star thistle	–	–	–	x
	Cichorium intybus	Chicory	x	–	–	–
	Cirsium arvense	Canada thistle	x	x	–	–
	C. vulgare	Bull thistle	x	x	x	x
	Conyza canadensis*	Horse weed	x	–	–	x
	Erigeron philadelphicus*	Fleabane	x	–	–	–
	Helianthus annuus*	Sunflower	x	–	–	x
	H. petiolaris*	Prairie sunflower	x	–	x	–
	Iva xanthifolia*	Marsh elder, sumpweed	x	–	x	–
	Krigia biflora*	Dwarf dandelion	x	–	–	–
	Lactuca canadensis*	Canadian lettuce	x	–	–	–
	L. serriola	Prickly lettuce	x	x	x	x
	L. tatarica*	Chickory lettuce	x	–	–	–
	Onopordum acanthium	Scotch thistle	–	–	–	x
	Sonchus asper	Prickly sow-thistle	x	–	–	–

Appendix A (continued)

	Scientific Name	Common Name	Jemez	Mesa Verde	Wupatki	Chuskas
Asteraceae (cont.)	Taraxacum officinale	Common dandelion	x	–	x	–
	Tragopogon dubius	Yellow salsify	x	–	x	–
	T. pratensis	Meadow salsify	x	–	–	x
	Verbesina encelioides*	Golden crownbeard	x	–	x	–
	Xanthium strumarium*	Cocklebur	x	–	x	x
Boraginaceae	Lappula occidentalis*	Flatspine stickseed	x	–	x	–
Brassicaceae	Alyssum minus	Yellow alyssum	–	x	–	–
	Brassica nigra	Black mustard	x	–	–	x
	Capsella bursa-pastoris	Shepherd's purse	x	–	–	x
	Cardaria draba	Hoary cress, white top	x	–	–	x
	Chorispora tenella	Blue mustard	–	–	–	x
	Cynoglossum officinale	Houndstongue, gypsy flower	–	x	–	–
	Descurainia sophia	Flaxweed tansy	x	x	–	x
	Lepidium latifolium	Perennial pepperweed	x	x	–	–
	Rorippa nasturtium-aquaticum*	Watercress	x	–	–	–
	Sisymbrium altissimum	Tumble mustard	x	–	x	–
	S. irio	Yellow rocket	x	–	–	–
	Thlaspi arvense	Field pennycress	x	–	–	–
Cannabaceae	Cannabis sativa	Marijuana, hemp	x	–	–	–
Caryophyllaceae	Cerastium arvense*	Meadow chickweed	x	–	–	–
	C. fontanum ssp. vulgare	Big chickweed	x	–	–	–
	Saponaria officinalis	Soapword, bouncing bet	x	–	–	–
	Silene antirrhina*	Sleepy catchfly	x	–	–	–
Chenopodiaceae	Chenopodium album*	Lambsquarters	x	–	x	–
	C. ambrosioides	Mexican tea	x	–	–	–
	C. desiccatum*	Goosefoot	x	–	–	–
	C. glaucum	Oakleaf goosefoot	x	–	–	–
	C. rubrum*	Red goosefoot	x	–	–	–
	Halogeton glomeratus	Saltlover	–	–	x	–
	Kochia scoparia	Mexican fireweed	x	–	x	–
	Salsola kali	Russian thistle	x	–	x	x
	S. tragus	Prickly Russian thistle	–	–	–	x

Appendix A (continued)

Family	Scientific Name	Common Name	Jemez	Mesa Verde	Wupatki	Chuskas
Clusiaceae	Hypericum perforatum	St. John's wort	x	–	–	–
Convolvulaceae	Convolvulus arvensis	Field bindweed	x	–	x	x
	Ipomoea hederacea	Ivy leaf morning glory	x	–	–	–
Euphorbiaceae	Croton texensis*	Doveweed	x	–	x	–
Fabaceae	Alhagi maurorum	Camelthorn	–	–	x	–
	Dalea leporina*	Foxtail prairie clover	x	–	–	–
	Medicago lupulina	Black medic	x	–	–	–
	M. polymorpha	Burclover	–	–	–	x
	M. sativa	Alfalfa	x	–	–	–
	Melilotus alba	White sweet clover	x	–	x	x
	M. officinalis	Yellow sweet clover	x	–	x	x
	Oxytropis lambertii*	Purple locoweed	x	–	–	–
	Trifolium hybridum	Alsike clover	x	–	–	–
	T. lappaceum	Burdock	–	x	–	–
	T. pretense	Red clover	x	–	–	–
	T. repens	White clover	x	–	–	–
Geraniaceae	Erodium cicutarium	Heron's bill, filaree	x	–	–	x
Lamiaceae	Marrubium vulgare	Common horehound	x	–	x	x
	Mentha arvensis*	Common mint	x	–	–	–
	M. spica	Spearmint	x	–	–	–
	Prunella vulgaris*	Healall	x	–	–	–
	Salvia pratensis	Introduced sage	x	–	–	–
Lemnaceae	Lemna sp.	Duckweed	x	–	–	–
Liliaceae	Asparagus officinalis	Asparagus	x	–	–	–
Malvaceae	Alcea rosea	Hollyhock	x	–	–	–
	Malva neglecta	Common mallow	x	–	–	–
Plantaginaceae	Plantago major*	Plantain	x	–	–	–
Poaceae	Aegilops cylindrica	Jointed goatgrass	x	–	–	–
	Agropyron cristatum	Crested wheatgrass	x	–	–	–
	Agrostis gigantea	Redtop	x	–	–	–
	A. scabra*	Rough bent grass	x	–	–	–
	Alopecurus aequalis*	Short-awn foxtail	x	–	–	–

Appendix A (continued)

Scientific Name	Common Name	Jemez	Mesa Verde	Wupatki	Chuskas
Poaceae (cont.)					
Aristida adscensionis*	Six-weeks three-awn	x	–	–	–
Beckmannia syzigachne*	Sloughgrass	x	–	–	–
Bromus inermis*	Smooth brome	x	x	–	–
B. japonicus	Japanese brome	x	–	–	–
B. rubens	Red brome	–	–	–	x
B. tectorum	Cheatgrass, downy chess	x	x	x	x
Cenchrus spinifex*	Sandbur	x	–	–	x
Chloris virgata*	Fingergrass	–	–	–	x
Cynodon dactylon	Bermuda grass	x	–	–	x
Cynosurus echinatus	Awned dogtail	x	–	–	–
Dactylis glomerata	Orchard grass	x	–	x	–
Dicanthelium xscoparioides*	Panic grass	x	–	–	–
Echinochloa crus-galli	Barnyard grass	x	–	–	x
Elymus repens	Quackgrass	x	–	–	–
Eragrostis cilianensis	Stinkgrass	x	–	–	–
E. curvula	Weeping lovegrass	x	–	–	–
E. hypnoides*	Teal lovegrass	x	–	–	–
E. pectinacea*	Tufted lovegrass	x	–	–	–
Festuca ovina	Sheep fescue	x	–	–	–
Hordeum jubatum*	Foxtail barley	x	–	x	–
H. murinum ssp. glaucum	Smooth barley	–	–	–	x
Leersia oryzoides*	Rice cutgrass	x	–	–	–
Lolium perenne	Ryegrass	x	–	–	–
L. pratense	Meadow ryegrass	x	x	–	–
Monroa squarrosa*	False buffalograss	x	–	–	–
Muhlenbergia rigens*	Deergrass	x	–	x	–
Panicum capillare*	Witchgrass	x	–	–	–
P. obtusum*	Vine mesquite	x	–	–	–
Pennisetum glaucum	Pearl millet	x	–	–	–
Phleum pratense	Timothy	x	–	–	–
Phragmites australis	Common reed	x	–	x	x
Poa annua	Annual bluegrass	x	–	–	–

Appendix A (continued)

	Scientific Name	Common Name	Jemez	Mesa Verde	Wupatki	Chuskas
Poaceae (cont.)	P. compressa	Canada bluegrass	x	–	–	–
	P. palustris*	Fowl bluegrass	x	–	–	–
	P. pratensis*	Kentucky bluegrass	x	x	–	–
	Polypogon monspeliensis	Rabbitfoot grass	x	–	–	x
	P. viridis	Rabbitfoot grass	–	–	–	x
	Setaria viridis	Green bristlegrass	x	–	–	–
	Sporobolus compositus*	Dropseed	x	–	–	–
	Thinopyrum intermedium	Intermediate wheatgrass	–	x	–	–
	Vulpia octoflora*	Six-weeks fescue	x	–	–	–
Polygonaceae	Polygonum aviculare	Knotweed	x	–	–	–
	P. convolvulus	Cornbind, black bindweed	x	–	–	–
	P. persicaria	Heart's ease	x	–	–	–
	Rumex acetosella	Red sheep sorrel	x	–	–	–
	R. crispus	Curlyleaf dock	x	–	–	x
	R. patientia	Patient dock	x	–	–	–
Portulacaceae	Portulaca oleracea*	Common purslane	x	–	–	x
Ranunculaceae	Ceratocephala testiculata	Bur buttercup, curveseed	–	x	–	–
Rosaceae	Potentilla norvegica*	Rough cinquefoil	x	–	–	–
Rubiaceae	Galium aparine*	Bedstraw, stickywilly	x	–	–	–
Scrophulariaceae	Linaria vulgaris	Butter & eggs	x	x	–	–
	Verbascum thapsus	Mullein	x	–	x	–
	Veronica anagallis-aquatica*	Water speedwell	x	–	–	–
	V. serpyllifolia*	Thymeleaf speedwell	x	–	–	–
Solanaceae	Hyoscyamus niger	Black henbane	x	–	–	–
	Solanum physalifolium*	Nightshade	x	–	–	–
	S. elaeagnifolium	Silverleaf nightshade	–	–	x	x
Urticaceae	Urtica dioica*	Stinging nettle	x	–	–	–
Zygophyllaceae	Tribulus terrestris	Puncture vine, goadhead	x	–	x	x

Appendix B

RARE, THREATENED, AND ENDANGERED PLANTS
FROM OUR STUDY AREAS

Appendix B. Rare, threatened, and endangered plants of our study areas.

Study Area	Scientific Name		Comments
Mesa Verde	Aceraceae	*Acer grandidentatum*	Critically imperiled in CO
Mesa Verde	Adiantaceae	*Adiantum aleuticum*	Imperiled in CO due to rarity
Mesa Verde	Apiaceae	*Aletes macdougalii breviradiatus*	Vulnerable globally; imperiled in CO due to rarity
Wupakti	Apiaceae	*Cymopterus megacephalus*	Endemic to Navajo Reservation; ranking G3, N3, Arizona S3, Navajo Nation S3 (vulnerable)
Jemez Mtns.	Apiaceae	*Oreoxis alpina* ssp. *alpine*	—
Wupatki	Apocynaceae	*Amsonia peeblesii*	Endemic occupying a narrow range; USFS listed as sensitive; BLM listed as sensitive candidate; ranking G3, N2, Arizona S3; Navajo Nation S1
Mesa Verde	Aspidaceae	*Dryopteris filix-mas*	Relic in Mesa Verde NP
Mesa Verde	Asteraceae	*Ambrosia tomentosa*	Critically imperiled in CO
Mesa Verde	Asteraceae	*Stephanomeria pauciflora*	Imperiled in CO due to rarity
Mesa Verde	Asteraceae	*Wyethia scabra*	Imperiled in CO due to rarity
Mesa Verde	Betulaceae	*Betula fontinalis*	Plant community of concern, CO
Wupatki	Bignoniaceae	*Chilopsis linearis*	Wetland indicator; ranking G5, N4N5, Arizona SNR
Mesa Verde	Boraginaceae	*Hackelia gracilenta*	Globally (and in CO); Mesa Verde endemic imperiled due to rarity
Jemez Mtns.	Boraginaceae	*Hackelia hirsuta*	NM uncommon species
Wupakti	Cactaceae	*Pediocactus peeblesianus* var. *fickeiseniae*	Critically imperiled; endemic to narrow range; Cat. 1 candidate for ESA listing; species of concern; ranking G1, G2, T1, T2; AZ S1, S2, Navajo Nation S1, USFWS R2
Wupatki	Cactaceae	*Pediocactus simpsonii*	Critically imperiled in AZ; ranking G4, N4, Arizona S1
Mesa Verde	Cactaceae	*Sclerocactus mesae-verdae*	Globally (and in CO) imperiled due to rarity
Jemez Mtns.	Cactaceae	*Sclerocactus papyrcanthus*	—
Jemez Mtns.	Campulanaceae	*Lobelia cardinalis*	Wetland species
Jemez Mtns.	Caryophyllaceae	*Silene plankii*	NM species of concern
Mesa Verde	Chenopodiaceae	*Suaeda moquinii*	Critically imperiled in CO
Jemez Mtns.	Cornaceae	*Cornus canadensis*	Wetland species
Mesa Verde	Crossosomataceae	*Forsellesia meionandra*	Critically imperiled in CO

Appendix B (continued)

Study Area		Scientific Name	Comments
Mesa Verde	Euphorbiaceae	*Chamaesyce fendleri*	Critically imperiled in CO
Wupatki	Euphorbiaceae	*Euphorbia exstipulata*	Ranking G5, NNR, Arizona SNR
Mesa Verde	Fabaceae	*Astragalus deterior*	Imperiled in CO due to rarity; CO endemic
Wupakti	Fabaceae	*Astragalus episcopus* var. *lancearius*	Imperiled in AZ and UT; ranking G3, G4, T2, T3, T2, N2, N3; Arizona S2, Utah S2S3
Mesa Verde	Fabaceae	*Astragalus humillimus*	Imperiled in CO due to rarity
Mesa Verde	Fabaceae	*Astragalus schmolliae*	Critically imperiled in CO; Mesa Verde endemic
Wupatki	Fabaceae	*Errazurizia rotundata*	Imperiled/rare in AZ; endemic to narrow range; ranking G2, N2, Arizona S2, Navajo Nation S1
Wupatki	Fabaceae	*Astragalus lentinginosus* var. *ambiguous*	Rare endemic, ranking G5, T1Q, N1; Arizona SU
Wupatki	Fabaceae	*Phaseolus angustissimus*	Ranking G4, NNR, Arizona SNR
Wupakti	Fabaceae	*Psorothamnus thompsoniae* var. *whitingii*	Navajo basin endemic; ranking T2, N2, Arizona S1; Navajo Nation S2
Wupakti	Fabaceae	*Senna bauhinioides*	Ranking G4, NNR, Arizona SNR
Mesa Verde	Fabaceae	*Trifolium gymnocarpum*	Critically imperiled in CO
Mesa Verde	Fagaceae	*Quercus ajoensis*	Critically imperiled in CO
Mesa Verde	Helleboraceae	*Aquilegia mancosana*	Imperiled globally (and CO) due to rarity
Wupatki	Hydrophyllaceae	*Phacelia serrata*	"Species of concern" by USFWS; endemic to volcanic cinders; ranking G3,N2, AZ S3
Mesa Verde	Frankeniaceae	*Frankenia jamesii*	Imperiled in CO due to rarity
Wupatki	Grossulariaceae	*Ribes cereum*	Noted "rare" by collectors; wetland indicator species
Wupatki	Hydrophyllaceae	*Phacelia welshii*	Imperiled in AZ, critically imperiled in NN; only 10 known occurrences; ranking G2, N2, AZ S2; NN S1
Mesa Verde, Wupatki	Lamiaceae	*Hedeoma drummondii*	Critically imperiled in CO; ranking G5, N5, Arizona SNR, Colorado SNR
Wupatki	Lamiaceae	*Hedeoma nana*	Ranking G5, N5, Arizona S4S5
Mesa Verde	Lamiaceae	*Monarda fistulosa*	Critically imperiled in CO

Study Area	Scientific Name		Comments
Sunset Crater	Lamiaceae	*Monarda pectinata*	Ranking G5
Wupakti	Lamiaceae	*Poliomintha incana*	Ranking G5, NNR, Arizona SNR
Mesa Verde	Liliaceae	*Fritillaria atropurpurea*	Imperiled in CO due to rarity
Jemez Mtns.	Liliaceae	*Lilium philadelphicum*	Wetland species
Jemez Mtns.	Loasaceae	*Mentzelia springeri*	NM species of concern
Mesa Verde	Malvaceae	*Iliamna grandiflora*	Imperiled globally (and CO) due to rarity
Mesa Verde	Malvaceae	*Sidalcea candida*	Critically imperiled in CO
Mesa Verde	Malvaceae	*Sidalcea neomexicana*	Critically imperiled in CO
Wupatki	Malvaceae	*Sphaeralcea leptophylla*	Ranking G5, NNR, Arizona SNR
Wupatki	Nyctaginaceae	*Allionia incarnata*	Ranking G5, NNR, Arizona SNR
Mesa Verde	Oleaceae	*Fraxinus anomala*	Imperiled in CO due to rarity
Wupatki	Onagraceae	*Camissonia specuicola*	Critically imperiled and endemic to AZ; ranking G2T1, N1, Arizona S1; Navajo Nation SNR
Wupakti	Onagraceae	*Gaura coccinea*	Ranking G5, N3N5; Arizona SNR
Wupakti	Onagraceae	*Camissonia boothii* ssp. *boothii*	Ranking T4, NNR, Arizona SNR
Mesa Verde	Orchidaceae	*Calypso bulbosa*	Critically imperiled in CO
Jemez Mtns.	Orchidaceae	*Cypripedium pubescens* var. *pubescens*	Wetland species
Mesa Verde, Jemez Mtns.	Orchidaceae	*Epipactis gigantean*	Imperiled in CO due to rarity; wetland species
Mesa Verde	Orchidaceae	*Spiranthes diluvialis*	Imperiled globally due to rarity
Sunset Crater	Pinaceae	*Pinus flexilis*	State protected in NV; ranking G5, N5, Arizona SNR
Mesa Verde	Pinaceae	*Pinus strobiformis*	Critically imperiled in CO
Wupatki	Poaceae	*Bouteloua barbata*	Ranking G5, NNR, Arizona SNR
Jemez Mtns., Wupatki	Poaceae	*Puccinellia parishii*	NM species of concern, imperiled in AZ; occupies specific fragile habitat; wetland indicator species; USFWS species of concern; ranking G2, N2, Arizona S2, Colorado S1, Navajo Nation S1, New Mexico S1
Jemez Mtns.	Poaceae	*Oryzopsis pungens*	Wetland species, threatened and endangered in several eastern states

Appendix B (continued)

Study Area	Scientific Name		Comments
Wupatki	Poaceae	*Sporobolus giganteus*	Wetland indictor species; ranking G5, NNR, Arizona SNR
Mesa Verde	Polemoniaceae	*Collomia grandiflora*	Imperiled in CO due to rarity
Mesa Verde	Polemoniaceae	*Ipomopsis gunnisonii*	Critically imperiled in CO
Mesa Verde	Portulacaceae	*Talinum parviflorum*	Imperiled in CO due to rarity
Mesa Verde	Primulaceae	*Primula specuicola*	Globally imperiled due to rarity
Jemez Mtns.	Ranunculaceae	*Delphinium sapellonis*	NM species of concern
Jemez Mtns.	Ranunculaceae	*Delphinium robustum*	NM sensitive species
Jemez Mtns.	Rosaceae	*Crataegus erythropoda*	Wetland species
Wupatki	Rosaceae	*Petrophyton caespitosum*	Ranking G5, N4, Arizona SNR
Jemez Mtns.	Salicaceae	*Salix arizonica*	Wetland species
Sunset Crater	Salicaceae	*Salix scouleriana*	Wetland indicator species; ranking G5, Arizona SNR
Mesa Verde	Scrophulariaceae	*Penstemon breviculus*	Vulnerable globally, imperiled in CO due to rarity
Wupatki	Scrophulariaceae	*Penstemon clutei*	Imperiled endemic in AZ; only 36 populationss known; fire dependent; ranking G2, N2, Arizona S2, Navajo Nation SNR
Mesa Verde	Scrophulariaceae	*Penstemon parviflorus*	Imperiled globally (and CO) due to rarity
Mesa Verde	Sinopteridaceae	*Pellaea glabellas* sp. *simplex*	Imperiled in CO due to rarity
Mesa Verde	Solanaceae	*Chamaesaracha coronupus*	Critically imperiled in CO
Mesa Verde	Typhaceae	*Typha angustifolia*	Critically imperiled in CO

Sources: Floyd and Colyer (2003); B. Jacobs (personal communication); Hogan and Huisinga (2000); NatureServe (www.natureserve.org/explorer); New Mexico Rare Plant Technical Council (http://nmrareplants.unm.edu/nmrptc/agency.htm); National Plant Data Center (http://plants.usda.gov).